TROUBLES IN THE RAINFOREST: British Columbia's Forest Economy in Transition

edited by

TREVOR J. BARNES & ROGER HAYTER

Canadian Western Geographical Series

Volume 33

Copyright 1997

Western Geographical Press

DEPARTMENT OF GEOGRAPHY, UNIVERSITY OF VICTORIA

P.O. BOX 3050, VICTORIA, BC, CANADA V8W 3P5

PHONE: (250)721-7331 FAX: (250)721-6216

EMAIL: FOSTER@UVVM.UVIC.CA

Canadian Western Geographical Series

editorial address

Harold D. Foster, Ph.D.
Department of Geography
University of Victoria
Victoria, British Columbia
Canada

Since publication began in 1970 the Western Geographical Series (now the Canadian and the International Western Geographical Series) has been generously supported by the Leon and Thea Koerner Foundation, the Social Science Federation of Canada, the National Centre for Atmospheric Research, the International Geographical Union Congress, the University of Victoria, the Natural Sciences Engineering Research Council of Canada, the Institute of the North American West, the University of Regina, the Potash and Phosphate Institute of Canada, and the Saskatchewan Agriculture and Food Department.

TROUBLES IN THE RAINFOREST

(Canadian western geographical series; 1203-1178; 33)
Includes bibliographical references.
ISBN 0-919838-23-5

1. Forests and forestry—British Columbia. 2. Forests and forestry—Economic aspects—British Columbia. I. Barnes, Trevor J. II. Hayter, Roger, 1947- III. Series.

SD146.B7T76 1997   333.75′09711   C97-910926-4

Series Editor's Acknowledgements

Several members of the Department of Geography, University of Victoria cooperated to ensure the successful publication of this volume of the Canadian Western Geographical Series. Special thanks are due to the technical services division. Diane Braithwaite undertook the very demanding task of typesetting, while cartography and cover design was in the expert hands of Ken Josephson. Their dedication and hard work is greatly appreciated.

The two paintings appearing on the cover, *B.C. Forest* and *Light Swooping Through*, are by Emily Carr (1871-1945). They are from the permanent collection of the Art Gallery of Greater Victoria, 1040 Moss Street, Victoria, British Columbia, V8V 4P1. Permission to reproduce these two works of art is gratefully acknowledged.

University of Victoria   Harold D. Foster
Victoria, British Columbia   *Series Editor*
November, 1997

Editors' Acknowledgements

This volume arose out of a conference of the same name held at both UBC and SFU in February, 1995. That event was itself the last of a year-long series of conferences organized across Canada under the auspices of Daniel Drache, Atkinson College, York University, and John Browne, Principal of Innis College, University of Toronto, to celebrate the centenary of the birth of the Canadian economic historian, Harold Adams Innis. We thank Daniel Drache for prompting the conference, and for his unflagging support and generosity. We also thank John Browne for his organizational skills, and Innis College for financial support especially towards the publication costs of this volume. In addition, Presidents Robert Pritchard, University of Toronto, and Susan Mann, York University, made monies available, as did the University of Toronto alumni who were students of Innis. The Harold Innis Research Institute and Innis College also aided the wider project, and our one in particular. Finally, we acknowledge a conference grant from SSHRC.

Closer to home we are grateful to the departments of geography both at the University of British Columbia and Simon Fraser University, and especially Tim Oke and Sandy Lapsky at UBC, and Dianne Sherry and Ray Squirrell at SFU. In addition, we would like to thank President John Stubbs, SFU, for attending the opening session and providing funds. Maureen Sioh and Robert and Tanya Behrish and Kevin Rees were wonderful, tireless assistants.

We also thank Carlyn Craig from BC Studies who helped us in a variety of ways, and the Special Collections Division, UBC Library, who assisted us in gathering the photos reprinted in this book. The photos originate in the MacMillan-Bloedel Photo Collection which is permanently stored at Special Collections. We also thank Special Collections for permission to reprint the photos here.

Four of the chapters of this volume have appeared in *BC Studies*, and we are grateful to the editors and publishers of that journal for allowing them to be reprinted here.

In addition, parts of Cole Harris's chapter are drawn from chapter six of his book *The Resettlement of British Columbia: Essays on colonialism and geographical change* (Vancouver, UBC Press, 1996). We would like to thank UBC Press for giving us permission to reprint sections of that essay.

Vancouver, British Columbia   Trevor J. Barnes
November, 1997   and Roger Hayter
                                                                        Editors

Contents

SECTION III: COMMUNITY IN TRANSITION

List of Figures

List of Tables

List of Plates

Plate 1 (overleaf)   Fallers at work   ➤

Troubles in the Rainforest: British Columbia's Forest Economy in Transition

Roger Hayter

Department of Geography, Simon Fraser University

Trevor J. Barnes

Department of Geography, University of British Columbia

Over the last decade and a half British Columbia's forests have become an increasingly troubled landscape. Continual environmental protests, aboriginal blockades, reduced annual allowable cuts, sweeping technological change, deep lay offs, changing global markets, multinational takeovers, and single industry communities scrambling for survival, have all shaken the once entrenched faith in British Columbia's "green gold".

The purpose of this book is to describe, analyse, and sometimes even suggest solutions to these troubles in the rainforest. The chapters are organized around three themes: the forest, the industry, and the community.

While each of the three raises a set of distinctive issues, each is also intertwined with the others. The scientific panel convened in 1993 to study Clayoquot Sound, for example, found that it was unable to examine only the ecosystem of the forest, its initial charge. Rather, it was necessarily led to consider the wider nature of the forest industry, principally represented by MacMillan Bloedel that held the rights to log in the area, and the communities that would be affected by any change including First Nations peoples, primarily the Nuu'Chah'Nulth band, as well as loggers, millworkers, and their families. The panel in effect recognized that the British Columbian forest sector should be conceived less as a system of individual components, and more as a tightly interlinked network of physical and social relations that shade into one another, so much so that the very distinction between the physical and social is sometimes hard to discern. For this reason the science of the forest ought to be as much social science as it is natural science.

The person who, better than anyone else, understood that Canadian resource industries are a dense complex of relations, rather than a set of disparate, separable components was the Canadian economic historian Harold Innis (1894-1952). Indeed, the conference on which this volume is based was one of several organized across the country during 1994-95 that celebrated the centenary of Innis's birth. Teaching at the University of Toronto for all of his academic life, Innis was a prolific writer and public intellectual. Perhaps his most famous thesis was staples theory, the idea that the economic history and geography of Canada could best be elucidated in terms of the export of various resource commodities such as fur, or fish, or, most germane to our case, forest products. While some have criticized the turgidness of Innis's theorizing, a more charitable interpretation is that he was struggling to resolve the immensely difficult problem of integrating into a single account the diverse relations existing within resource industries: physical geography, institutions, markets, transportation, single industry communities, and a core-periphery regional structure. This book's structure, which stresses the interconnectedness of forests, industry, and community, strives to be part of this same Innisian tradition by emphasizing synthesis and integration.

FORESTS IN TRANSITION

The forest as a physical environment is at the heart of British Columbia's forest economy. But it is a physical environment that belies any final, objective assessment (Willems-Braun, 1996). Rather, the forest means many different things to different people, not all of which are reducible to one another: it is profit and jobs, it is an awe-inspiring vista, it is a complex ecological unit, it is also "Mother Earth", an inspiration of myths, and a place called home. These are not mere semantic differences of interpretation. Each of these views takes on specific material forms and practices that often conflict with the other interpretations: clear cutting clashes with awe-inspiring vistas, and commercial mass tourism is incompatible with an eco-tourism that reveres pristine nature. In each case, something needs to give, and what gives often creates friction. In the past, that friction was often limited and contained by powerful commercial interests in alliance with the local state, the provincial government. That no longer seems possible. An increasingly active environmental movement (Greenpeace was founded in Vancouver in 1969), widening political mobilization of First Nations peoples (Blomley, 1996), the decline in power of forest companies themselves, and the recognition by the local state of the legitimacy of different forestry uses, have made competing interpretations of the forest environment, and hence disputes around it, much more likely and visible.
        Historically, the disputes over the forest are relatively recent. While aboriginal peoples have drawn on the resources of the cedar-fir-hemlock British Columbia coastal forest ever since first settlement, roughly 10,000 years ago,

the first systematic commercial exploitation of British Columbia's temperate rainforest did not begin until the 1850s. Such exploitation represented the western-most extension of European land alienation and resource appropriation that began in New Brunswick in the seventeenth century, and which later continued through Quebec and Ontario. During much of this formative period, the "assault" (Lower, 1938) on Canadian forests was carried out through a strategy of "cut and run", thereby progressively eliminating forests as they became accessible. This same tactic of forest "liquidation" (Lower, 1938) was also pursued during the early years of commercial harvesting of forests in British Columbia. Such was the speed and extent of the cutting that by the late 1880s the provincial government introduced timber leases. Even then, however, the primary goal was less to conserve the forest than to ensure that outside investors would generate employment and capital for British Columbia.

During the immediate postwar years it was the Sloan commission of 1947, and a subsequent one in 1956, that shaped British Columbia's policy of forest use and control. Two main types of tenure were established, the tree farm license (TFL) and the public sustained yield unit (PSYU), both of which comprised large tracts of forest which were to be leased to private companies following a tendering process that required potential lessees to submit plans for forest use. Because of their stability, resources, and long-term horizons, the Ministry of Forests typically chose bids from larger forest companies over smaller ones, and in this sense, the implicit agenda of the Sloan Commissions favoured big business.

In retrospect, it is clear that the provincial government's primary purpose during this period was ensuring high levels of investment and employment, and not conserving the forest resource. Certainly, maximizing stumpage rates, the resource royalty charged companies for harvesting on Crown land, and managing the forest stock, were not priorities. Because the government feared that high stumpage rates might dissuade investment, low stumpage levels were preferred; in fact, until the early 1980s, the collection costs of stumpage exceeded the revenue generated (Travers, 1993: 197-204). Likewise, while Justice Sloan had introduced the notion of sustainable yield, there was no real systematic attempt to apply that notion. At best there was the generally held, and rarely challenged, belief that natural regeneration would adequately restock the forests in time for the next cycle of harvesting. Until the 1970s, then, with the possible exception of some privately held lands, neither the private nor public sector acted as even modestly responsible stewards for the resource, with negligible tree planting and silviculture.

The first signs of change appeared in the early 1970s. According to the government's own calculations, there was increasingly little slack between allowable and actual harvesting levels on the province's coast. This was interpreted as a consequence of the so-called "fall-down" effect, that is, a "natural" decline in forest productivity as industry moves from harvesting old growth to secondary forests. Environmentalists argued, however, that "fall down" was a symptom of both the historic neglect of forest management in the province, and, more pressingly, indicated the preciously small amounts of old growth forests remaining, the ecological characteristics and values of which could never be regained once harvested. The stage was thus set for a series of valley-by-valley, island-by-island protests, most notably the Waldren, Carmanagh, the Stein, Meares Island, and the Queen Charlottes, culminating in the protest to end all protests, mass civil disobedience at Clayoquot Sound in the summer of 1993.

Sometimes allied with these environmental protests, but also a powerful independent voice, are First Nations peoples. With no land treaties having ever been signed with the Province of British Columbia, aboriginal peoples had negligble property rights or control over resources. While there were some protests by First Nations Peoples in the early and mid-1970s, forest companies and the lands that they logged were not systematically targeted until a decade later. At that point there were a series of confrontations and blockades. In 1984, for example, the Nuu'chah'nulth blocked logging access to Meares Island, while a year later the Haida obstructed logging on Lyell Island (Tennant, 1990: 207; also see Blomley, 1996). The reason for this change in strategy was that by the mid-1980s many of ". . . the protesting groups had prepared and submitted land claims [to the Federal government], and they were now seeking to protect the lands and the resources until the claim was settled" (Tennant, 1990: 207).

While both environmental and First Nations disputes of the 1970s and 1980s continued into the 1990s, the election in 1991 of the B.C. New Democratic Party presaged significant alterations to postwar forest policy in British Columbia. Following a series of Round Table discussions around economy and environment, and the establishment of a Commission On Resources and the Environment (CORE), the NDP government directly tackled a series of highly contentious issues that previous governments had either ignored or stubbornly resisted resolving. First, with respect to land use, Stephen Owen working under the umbrella of CORE was charged with preparing, in consultation with local community groups, detailed regional plans for land and resource use (plans have been drawn up for Vancouver Island, the Kootenays, and the Cariboo-Chilcotin). Second, with respect to forest practices, a strict Forest practices code was passed in the legislature, and a new crown corporation, B.C. Forest Renewal, was initiated and funded by higher stumpage rates all with the wider purpose of improving forestry management. Third, with respect to aboriginal claims, in February 1996 the provincial government signed an agreement in principle with the Nisga'a band, which was the first of the land claims submitted to the Federal government back in 1974. Finally, with respect to

Clayoquot Sound, a scientific panel was formed to investigate the local ecosystem of the area and to consider the effects of logging, and reported its findings which the government accepted in the autumn of 1995.

Of course, it is possible to interpret these recent changes as only cosmetic, of papering over cracks that will inevitably reappear again once, say, there is change of government. A more optimistic reading, though, is that for the first time since European settlement there is the hope of a more pluralistic, less hegemonic, conception of the forests; one in which multiple interpretations can exist side-by-side, not necessarily in harmony but at least starting to respect one another. It might not be a "Forestopia" (M'Gonigle and Parfitt, 1994), but it need not be "war in the woods" either.

The first section of the book is devoted to the complexities and often fractious nature of these recent transitions in the forest. Clark Binkley's essay begins by taking stock of current provincial land use policy, and concludes with some radical proposals for improvement. Michael M'Gonigle also offers radical proposals for improvement, but he takes a very different environmental position, arguing that the province should organize land use to minimize harvesting and not to maximize it as in Binkley's plan. Bill Cafferata's essay stays away from these big environmental issues, offering instead a very detailed empirical assessment of British Columbia's forest resource base and the effect on it of current logging practices. Both Tom Gunton's and Jeremy Wilson's papers that follow examine the changing but also continuing pivotal role of government policy in setting the agenda for the various resource and land use conflicts recently dogging the industry. The last essay in the section, by Bruce Willems-Braun, situates British Columbia aboriginal land claims and the forest industry within a framework based upon recent writings in cultural studies and around the politics of nature.

INDUSTRY IN TRANSITION

A central player in all of the "troubles" has been the forest industry. Given the plethora of criticisms levelled at it, and its various trials and tribulations, it is easy to forget that the wood products sector remains overwhelmingly the province's most important manufacturing industry, generating, directly and indirectly, significant employment, value added, and export and tax revenue. While the old rule of thumb that the sector comprised 50 percent of the provincial economy now clearly overstates the case, even conservative estimates indicate that in the late 1980s the forest sector accounted for about 17 percent of provincial income (taking into account indirect effects) and over 14 percent of provincial employment (Travers, 1993: 183). In terms of exports, forest products continue to define the province's global role.

That said, the industry experienced a series of profound shocks over the last 15 years, the severity of which led some to say that the forestry sector is now in its sunset years. However, the industry has always been in flux, featuring cycles of boom and bust. But for the 30 year period after the end of the Second World War it experienced relative stability. Under a regime of Fordism, one defined by an implicit coalition among corporate capital, organized labour, and the local state, forest companies mass produced a limited number of bulk, standardized products, principally, lumber, kraft pulp, and plywood, using production-line technology, and which were then sold primarily to U.S. and Canadian markets.

Undergirding the success of Fordism in the province was the favourable access to forestry resources. As already suggested, the Sloan commission paved the way for large-scale industrial development in the province, the major players of which were big, often foreign-owned corporations (in 1970, 35 percent of lumber and 45 percent of pulp production was foreign owned). Aiding and abetting were the provincial government that provided necessary infrastructure, and organized labour that tacitly accepted mass production techniques and organization in return for steadily increasing wages (around 5 percent above the Canadian industry norm). The pulp and paper industry, in particular, was central. Between 1947-72 over 20 new capital-intensive, export-orientated pulp and paper mills were constructed. Prior to 1961 these mills were built on the coast, but after that date they were increasingly constructed in the interior. Over time the pulp and paper mills became more integrated into other aspects of the wood processing industry. For example, "waste" from sawmills such as wood chips and sawdust were used respectively as both fibre and fuel.

The relatively stable growth of the 1960s, however, became more turbulent during the 1970s. As questions were first being raised about the adequacy of the forest resource base, markets for low value wood products became more volatile and competitive. Likewise the pace of technological change also increased, creating heightened uncertainty for workers. This period culminated with the deep recession of the early 1980s which devastated the industry, and provided confirmatory evidence, if evidence was still needed, that the long boom was over. American multinational corporations started leaving the province in the mid-1970s, and British Columbia's own MacMillan Bloedel was sold and resold over the first part of the 1980s, and dramatically restructured in the process, shedding thousands of workers along the way.

Things were clearly not the way they had been. The British Columbia forest sector, and Western economies more generally, were entering a different, post-Fordist industrial regime. Defined by computerized production methods, increased globalization, specialized niche markets, and a dramatically reduced demand for un- and semi-skilled unionized labour, the coming of post-Fordism necessitated a fundamental restructuring of the forest industry, especially on the coast. Computerized production technology was introduced in mills, new markets, often Asian, were established, old products were discontinued and more specialized ones introduced, and everywhere there were permanent lay offs; the number of jobs directly related to the processing of 1,000 cubic metres of board lumber have more than halved since 1961, for example (Travers, 1993).

The 1980s was a troubling time for the industry, and it was further exacerbated by problems around the forest base itself, and by ongoing trade disputes with the U.S. over softwood lumber. But, as with the discord around the forest base, these recent changes in the forest industry are interpretable in two ways. The pessimistic interpretation is of an industry on the decline, fizzling out as the resource base itself disappears. The optimistic interpretation, though, is of a kind of industrial renaissance, of a newly fashioned forest products industry that emphasizes high-value products, skilled labour, and leading edge technology. It might not be the same as the old times, but the old times were not necessarily that good.

Examining both the old and the new times of the industry is the purpose of the second section of the book. The first two chapters hone in on different facets of the trend towards globalization. Bruce Wilkinson's paper is the most general, and documents Canada's and British Columbia's reliance on staples industries, particularly forestry, and the mutating geographies of their associated markets. Pat Marchak's focus is less the geography of markets, and more the geography of production. Her argument is that, with forestry production sites opening up world wide, particularly in developing countries where environmental standards are more lax compared to those of Canada, there is a switch of foreign investment away from British Columbia with dire consequences for provincial employment and wealth generation. Otto Forgac's chapter, in contrast, examines issues closer to home, specifically the role of research and development within the forestry sector, especially in the design of new wood products. Casting these latest research and development gains within a wider framework—post-Fordism—Hayter and Barnes in their chapter draw out the consequent implications for employment and community sustainability.

COMMUNITIES IN TRANSITION

The third section of the book begins where Hayter and Barnes's chapter ends, with issues of community stability. The single industry communities that dot the province are the underpinning for the production of British Columbia wood products. They are the sites where global forces and the actual extraction of the resource most immediately come together.

The essence of a single industry town is its reliance on the production of one resource commodity, which is often destined for export to foreign markets. In Innis's terms, resource communities are the basis of the hinterland, and their role is to provide staple commodities for the metropole (that is the core economic region). Whereas the hinterland engages in only low value-added upstream production, the metropole is where high value-added manufacture occurs, final products are distributed, and key decision-making about resource production takes place. Innis's dualistic spatial scheme that turns on the distinction between hinterland and metropole seems a perfect fit for British Columbia. Ever since the first part of the twentieth century Vancouver has served as the province's local metropole, and has been the site of distribution and decision-making for the staples production carried out in the rest of the province. In fact, until the 1950s British Columbia's resource hinterland led the nation in terms of the absolute number of people living in resource communities. Even now, by some estimates, a quarter of British Columbia's population is resident in single industry towns, there being at the last count over 100 resource communities.

The perennial problem with single industry communities is their instability, a result of the very economic nature of the resources on which they depend. Because of the vicissitudes of international commodity prices, international corporate decision-making, technological change, and the supply and quality of the natural resource stocks themselves, thriving single industry towns can quickly become ghost towns.

While the fortunes of single industry communities in British Columbia have waxed and waned historically, during the last 15 years they have mostly waned. Twenty-five years ago the archetypal British Columbia logging town, Port Alberni, was the eighth most prosperous community in Canada; now it is not even in the top 100. The causes of this change stem precisely from the problems already noted that revolve around industrial restructuring and forest use and management. The upshot is that since 1980 a number of mills, especially coastal, have been shut down, wound down, or reconfigured to varying degrees, creating a wide range of problems for the associated single industry communities.

Given these recent changes, it is not surprising that the issue of local economic policy has become paramount. It has taken quite a different form, however, from the traditional central government, top-down regional development strategies that were so prevalent in the 1960s. Because of both the realization that centralized policies have not worked well, and a new political climate of neoconservativism that stresses individual and community initiative, community development is now conceived as a bottom-up process in which localities and the private entrepreneur, acting singly or as a part of a locally-based coalition, are the primary agents of change. The consequence is that local development is now expressed in a wide variety of schemes in a wide variety of sectors. Whether these initiatives can offset the losses of employment occurring in the mills is still unclear.

While these new forms of community development were occurring in the hinterland, there were also corresponding changes to the nature of Vancouver as the local metropole. Increasingly integrated into a matrix of world cities on the edge of the Pacific Rim, Vancouver has, according to some commentators, "decoupled" itself from the staples hinterland that formerly provided its rationale. Some even speak of the emergence of a post-staples economy, one in which the external links forged by Vancouver are now more important than any internal provincial ones.

As with forests and with industry, the changes affecting British Columbia's hinterland and metropole over the last 15 years can be interpreted in different ways. From one perspective, they can be viewed as the unravelling of British Columbia's very way of life; as a case of the centre no longer holding. But from another, they are the harbinger of something better, not necessarily yet realized but in the offing; a future in which communities are no longer critically dependent upon the vagaries of markets and resource supplies, or the local metropole.

Cole Harris' essay that begins this section offers a broad historical view of the fate of single industry communities, and ties their vacillating fortunes to technological improvements in transportation and communication, which for him are the very embodiment of modernity. Tom Hutton's essay argues that keeping British Columbia's single industry communities in place has historically been the power and decision-making role of Vancouver. This is now changing, though, as Vancouver becomes a world city, and British Columbia becomes a "post-staples" economy. Maureen Reed and Alison Gill look at the flip side of that same process by drawing upon a case study of Squamish, and examining the kind of local economic development strategies single industry towns must now pursue in a new post-staples economy. Finally, Gerald Walter offers a very broad approach to community by melding together ideas of sustainable development and local economic development, using the work of Harold Innis as a starting point.

CONCLUSION

Throughout British Columbia, but especially on the coast, the structure of, and interrelationships among, forests, forest industries and forest communities are radically different in the latter part of the 1990s than they were even in the early 1980s. On the coast old growth is still harvested, but the shift towards secondary growth harvesting is now well under way and advancing rapidly.

Even more remarkably, hardwood plantations, such as MacMillan Bloedel's popular plantation in the Alberni Valley, have been established and will add to the supply of wood fibre in a short period of time. Forestry practices have long extolled principles of sustained yield and multiple use. But since 1980 these principles have been monitored and regulated more closely, and traditional industry values have been considerably modified by broader economic, environmental, and social concerns. Industrial forestry practices have become more ecologically sound and geographically restricted.

Faced with the prospect of less and more expensive fibre, and competition from lower cost regions elsewhere, the forest industry has shifted rapidly from a narrow reliance on a few bulk commodities to a more diversified, higher value range of products. Large scale operations remain a dominant feature of the industry, but increasingly mass production has been redefined as flexible mass production in which big plants are capable of producing a more differentiated range of products. Moreover, these new, highly automated big plants have been complemented by growth of a secondary wood industry organized around much smaller, more flexibly specialized firms. Shifts in trade patterns have been closely associated with this shift to flexibility as heavy dependence on the U.S. has been replaced by the rapid growth of Japanese markets.

Within the hinterland, especially on the coast, forest communities have been transformed by a combination of industry rationalization and modernization, shifts towards service sector employment, and rapid population dynamics. If communities such as Chemainus and Port Alberni are still forest communities in which the forest industry offers the "best" jobs, they are no longer "just" forest communities. In such communities, a declining proportion of the population has direct links to the forest industry and community attitudes and expectations are no longer inherently empathetic towards the big mills that for so long dominated the local economic base. While forest communities are increasingly attracting newcomers with no ties to the forest industry, perhaps the most telling transformation of community values is among high school students who no longer see their immediate or long term future in the forest industry.

The fact that forests, forest industries, and forest communities in British Columbia have changed radically should not belittle their continuing importance to the life of the province. Their influence on the character of British Columbia and its global role is both apparent and assured.

REFERENCES

Blomley, N. (1996). "Shut the province down": First Nation blockades in British Columbia, 1984-95. *B.C. Studies*, 111, 5-36.

Willems-Braun, B. (1996). *Materializing nature: Discourse, practice and power in the temperate rainforest.* Unpublished PhD dissertation, Department of Geography, University of British Columbia.

Lower, A.R.M. (1938). *The North American assault on the Canadian forest.* Toronto: Ryerson.

M'Gonigle, M., and Parfitt, B. (1994). *Forestopia: A practical guide to the new forest economy.* Madeira Park, B.C.: Harbour Publishing

Tennant, P. (1990). *Aboriginal people and politics: The Indian land question in British Columbia, 1849-1989.* Vancouver: UBC Press.

Travers, O.R. (1993). Forest policy: Rhetoric and reality. In K. Drushka, B. Nixon, and R. Travers (Eds.), *Touch wood: B.C. forests at the crossroads* (pp. 171-224). Madeira Park, B.C.: Harbour Publishing.

SECTION I
Forests in Transition

A Cross Road in the Forest: The Path to a Sustainable Forest Sector in British Columbia*

2

Clark S. Binkley

Faculty of Forestry, University of British Columbia

* This essay first appeared in *BC Studies*, Spring, 1997

> *I shall be telling this with a sigh*
> *Somewhere ages and ages hence:*
> *Two roads diverged in a wood, and I—*
> *I took the one less traveled by,*
> *And that has made all the difference*

> **The Road Not Taken**
> Robert Frost (1916)

British Columbia lies at a crossroads in the transition between forests provided by providence and those created through human husbandry and stewardship. Many of the changes now tormenting British Columbia are predictable consequences of human interaction with primeval forests. Indeed, the earliest recorded story—the *Epic of Gilgamesh* written in cuneiform on a clay tablet 5,000 years ago—remarks on the dire consequences of forest depletion. Each subsequent civilization has relived this story with little change in the theme, from the Greeks in the Mediterranean, to the wandering bands in Central Europe, the Swedes late in the last century and our southern neighbours in the last decade or so.

Large expanses of virgin forest remain in only a few places on earth—in British Columbia and elsewhere in Canada, in eastern Russia, in the Amazon, and in parts of Africa. Those in British Columbia lie on the cusp of an irreversible slide into the established historical pattern of resource depletion and attendant social disruption. But, unlike most other developed parts of the world, in British Columbia there still is an opportunity to make the changes needed to sustain a vast wild estate while continuing a prosperous society based on forest resources.

This chapter contemplates British Columbia's predicament in three parts. First, by way of background, what are the predictable elements of resource depletion as they have evolved in other parts of the world? Second, what are some of the key elements of a sustainable forest sector for the future? Finally, how must policies adjust to move from the present, uncomfortable position to a preferable future?

TIMBER DEPLETION AND FOREST SECTOR DEVELOPMENT

To examine the predictable links between timber depletion, forest sector development, and societal evolution, the economist's mode of analysis is used here: hold everything else constant to focus clearly on the issue at hand, with full knowledge that the world contains a much richer fabric of complications. In this spirit, consider a sovereign entity (called a "country" below) embedded in a world forest sector. Imagine the country discovers an old-growth timber resource large enough for its development to be significant in the relevant markets. In the absence of policy intervention, how does the forest sector develop?

The first part of this story is the comparatively old and well-known saga of the intertemporal adjustment of a timber stock (Lyon, 1981; Sedjo and Lyon, 1990; Sedjo, 1990). The second part links the dynamics of the timber stock to the changes in the forest sector, the macro-economy and the environment more broadly (Vincent and Binkley, 1992).

Dynamic Adjustment of the Timber Stock

In the early stages of development, net growth of the forest is nil: photosynthesis just balances the death of plant tissues and entire trees. Because growth is nil, any harvest at all exceeds the growth of the forest. Since harvest is greater than growth, the timber inventory declines.

As the inventory of old-growth timber declines, timber becomes scarcer. In economics, "scarcity" is a synonym for higher prices. Harvesting costs will increase as logging pushes into increasingly remote sites. Timber rents—the value of the standing timber itself—will increase as a consequence of old-growth depletion and the link between timber markets and capital markets. Prices rise until the purposeful husbandry of second-growth timber or the use of non-wood substitutes (stone, concrete or steel for construction; fossil fuels, solar energy and conservation for energy) becomes economic. Because timber prices have risen, other countries can compete in world markets, either by exploiting their own old-growth reserves (e.g., the Philippines, Indonesia, or Malaysia), or by developing plantations (e.g., New Zealand, Chile, Brazil, South Africa).[1]

In the absence of subsidies, high-cost old growth is apt to remain unharvested and become economic wilderness.[2] We see this today in remote locations of British Columbia and other parts of the world. What do these adjustments in the timber stock imply for such important concerns as capital, labour, and the environment?

Capital and Labour

In economic terms, harvesting timber transforms *ecological* capital into *economic* capital. The reduction in timber stocks increases the supply of capital to the economy. The price of capital will decline relative to the price of labour. Because timber prices are rising, the price of capital falls relative to the price of timber even more.

The technology used to process timber logically adapts to these changes in factor prices: firms substitute capital for labour and capital for timber. Because of the nature of the technologies used in the forest sector, capital-timber substitution typically reduces specific labour utilization as well. The resulting increases in technical efficiency push up output per person-hour (i.e., labour productivity) and open the door for higher wages. But the mathematical inverse of labour productivity is employment per unit of output, so efficiency gains, all else equal, mean fewer workers per unit of output. As a result of these effects, employment per cubic metre of timber harvested in British Columbia has fallen by a factor of two between 1961 and 1989 (Nixon, 1991).

An iron law of economics explains that the economic return to the addition of any one factor of production declines as more of that factor is used. As firms use more and more capital in an effort to offset rising relative prices of timber and labour, the single-factor productivity of capital will decline. Lower returns on capital mean less capital investment. Unless technical efficiency increases via the application of new technologies, wage rates ultimately must fall (or at least not rise as much as they do in other countries) or the forest sector will fall into a self-perpetuating spiral of declining productivity.

Increases in timber prices will drive up product prices. Higher product prices encourage substitution away from conventional wood-based products. So, for example, the consumption of softwood lumber in the United States has remained roughly constant since 1900 at 40 to 50 billion board feet annually, despite a sevenfold increase in economic activity and more than a doubling of the population in that country. Some of this substitution has been pure efficiency gain (e.g., because of improved knowledge about the performance of wood used in buildings, a basic framing member that measured 2" x 4" in cross-section at the turn of the century has now been reduced to 1.5" x 3.5" with no loss of building safety), some has been wood-wood substitution (e.g., in sheathing applications, first plywood substituted for lumber; more recently oriented strandboard has substituted for plywood), but some has replaced wood with other materials (steel studs, plastic bags, and concrete buildings).

Environmental Values

As per-capita income increases, so too will the demand for the services of natural environments. Empirical evidence substantiates this claim for some

features of the environment (e.g., air quality, water quality and outdoor recreation) and the positive relation between income and environmental values
is probably more broadly applicable as well.[3]

At the same time, development of forests for timber production will decrease the supply of these environmental services. Local and global environmental services operate outside formal markets, either because they are true
public goods (i.e., aesthetically pleasing landscapes, carbon sinks) or because
society has chosen not to allocate them through markets (clean water flowing
from a forested watershed, recreation). It is a simple truism that such goods
are systematically undervalued in forest consumption and production decisions. As a consequence, market-based patterns of forest use cannot and do
not reflect the social values of these inputs and outputs. This mismatch between the social valuation of natural environments and their value in formal
markets will widen over time. Once the gulf is sufficiently large to overcome
transactions costs, institutions will emerge to place values on these services.
They may actually be traded in formal markets, or they may be protected
through direct governmental ownership, through regulation or through elaborate and costly mechanisms of planning and public involvement in forest
decision making. Whatever the mechanism, the costs of these formerly free
environmental inputs will rise, and productivity as conventionally measured
will decline as these previously free inputs come to carry positive costs.

CHALLENGES FOR A SUSTAINABLE
FOREST SECTOR IN BRITISH COLUMBIA

The problems British Columbia faces today are no more or no less than the
local manifestation of the more general phenomena outlined above. In the
absence of positive adaptations in public policy, British Columbia faces wrench-
ing structural adjustments in its economy, community and regional patterns of
development. The well-known "fall down effect" in British Columbia—the
planned reduction in timber harvests as old growth is depleted and investment in second growth is inadequate—is but one example.

In 1992 the British Columbia government announced a review of long-
term timber supply on all Timber Supply Areas (TSAs) and Tree Farm Licences
(TFLs)—virtually the entire land base which supports industrial activities related to British Columbia's forest sector. Under the Forest Act, the level of
annual allowable cut (AAC) for individual areas is not calculated from any
one formula, but rather is set by the Chief Forester on the basis of broad biophysical and socio-economic criteria. As a result, it is nearly impossible to
predict future AACs. However, the analytical models used by the Ministry of
Forests to estimate potential future harvest levels give an indication of the results of current policy direction. At this writing, the Ministry has released

analytical reports for 25 of the 36 TSAs, representing some 35.3 million cubic metres. of AAC (versus about 72 million cubic metres for TSAs and TFLs province-wide). These reports collectively suggest that current policy regimes will result in about a 23.5 percent long-term reduction in provincial total harvest levels, with a significantly greater impact on the Coast than in the Interior (Miller, personal communication).[4]

What are the implications of this level of reduction in timber supply? A recent study (Binkley et al., 1994) examined a variety of economic impact analyses related to harvest reductions. It concluded that a 25 percent reduction in harvest levels will mean a loss of up to 92,000 jobs and 4.9 billion dollars in GDP in the province with more-than-proportional impacts on governmental revenues (i.e., the net loss of taxes on GDP grossed up by increases in social service costs for unemployed workers). Although even Vancouver's economy relies heavily on the forest sector,[5] the impacts would be felt most strongly in the 39 of 55 rural communities in British Columbia where the forest sector is the dominant basic industry (Ministry of Finance and Corporate Relations, 1992). These economic effects are likely to produce social effects that go along with community disruption (e.g., alcoholism, divorce, suicide). These social problems will come just at the time that the province is least able to assist: this study indicates that, with a 25 percent reduction in harvests, the provincial budget deficit will increase by about 2 billion dollars.

The obviously large, negative impacts of current policy direction suggest that significant benefits might accrue from a change in direction. We are indeed at a cross roads in the development of our forest sector. Before turning to desirable policy changes—the path best taken—let us examine some desirable policy outcomes. These include simultaneously maintaining harvest levels, enhancing the productivity of the forest sector at high wage rates within the constraints of available timber, and sustaining important ecological and environmental characteristics of British Columbia's diverse forest estate.

Maintaining Harvest Levels

Current policies are apt to lead to significant reductions in harvest levels. Are these reductions necessary?

Historically, British Columbia's forests have been managed extensively under the implicit assumption that virtually the whole forested land base would, one day, be available for timber production. The B.C. Forest Service and licensees incorporated non-timber values into the timber production plans through a process of "integrated resource management" which attempted to consider wildlife, riparian habitat, recreation, water flows, grazing and other forest uses on each hectare where logging was to occur. Investments in silviculture were low. While licensees are now required to regenerate all areas logged to a "free to grow" stage and massive reforestation efforts under the various federal/provincial agreements have virtually eliminated the backlog of "not satisfactorily re-stocked" (NSR) lands, British Columbia's use of silvicultural technology lags behind that in virtually every country with which the province competes.

This approach to land management has clearly failed. It does not satisfy those concerned with the non-timber values of the forests. It is a clear prescription for reduced harvest levels, and responds neither to the commercial needs of the forest sector, nor the economic needs of communities, nor the financial needs of the provincial government. Recent theoretical work (Vincent and Binkley, 1994; Swallow et al., 1990) and empirical analysis (Sahajanathan, 1994)[6] confirms the old idea that the multiple benefits of forests are best provided by zoning the forest into a series of special-use areas corresponding to the range of forest values society demands—from wilderness to timber production. The vanguard of the environmental movement understands the wisdom of this approach (e.g., McNeely, 1993; Alverson et al., 1994).

British Columbia has moved tentatively in this direction through the process established by the Commission on Resources and the Environment (CORE). In the three areas studied (Vancouver Island, the Cariboo-Chilcotin, and the Kootenays), CORE defined zones ranging from protected areas to areas for intensive timber production. The details of management in each zone remain to be developed, but if the management rules actually do permit intensive management on some of the land, then it may be possible to offset much of the planned reduction in harvest levels.

Consider a benchmark. In the late 1970s, the Weyerhaeuser Company studied the biochemical efficiency of trees in turning sunlight into wood, and modelled maximum biologically possible timber yields (Farnum et al., 1983). They applied this model to two sites in the United States—one in the Pacific Northwest (Douglas fir), and a second in the Southeast (loblolly pine)—where they practice as intensive forestry as can be found anywhere in the world. For example, in the Southeast, the study plantations were site-prepared, bedded, fertilized, planted with genetically improved/mycorrhizal-inoculated seedlings, optimally spaced after planting and repeatedly thinned and fertilized. Yet the production of these stands achieved only 40 to 50 percent of the theoretical yields. Natural stands in the same locations grow only about 10 to 25 percent of the theoretical yields.

Forests in British Columbia are managed much less intensively than are these study sites. As a consequence, there appears to be considerable latitude for increasing the production of economically usable plant parts in the province. With intensive silviculture, yields of from two to five times the levels attainable in natural stands generally appear to be economic.

There is some local empirical evidence that these kinds of increased yields are indeed feasible in British Columbia. Tolnai's (1991) analysis of Weyerhaeuser's TFL 35 near Kamloops found that, with more intensive management, harvest levels on this licence could be sustainably increased by 70.1 percent. In some stands the increase was even more dramatic, rising from 2.3 cubic metres per hectare per year under the current management regime to 8.3 cubic metres per hectare per year in a more intensive management regime. And, remarkably, these management regimes involved only prompt restocking with desirable species (in this case, lodgepole pine) at an appropriate density. Increased yields from genetic improvement of planting stock and site emendations could be added to these gains.

The impediments to achieving these gains are mainly institutional. Licensees have no incentive to invest in more intensive silviculture because the gains accrue to the Crown and not to them. From a financial point of view, private investments in silviculture are equivalent to tearing up money and throwing it out the boardroom window: they neither enter the balance sheet, nor produce incremental earnings in the future.

Enhancing Productivity

Economists define "productivity" simply as the value of an industry's outputs divided by the costs of its inputs. Increases in productivity are obviously prerequisite to increases in material standards of living. Porter (1990) expresses the situation well:

> *Productivity is the prime determinant in the long run of a nation's standard of living, for it is the root cause of national per capita income.*

In his study of Canada, he goes on to say

> *Canada's economy, and especially its export economy, is heavily based on natural resources. Some argue that resource industries are inherently less desirable than manufacturing or "high tech" industries. This logic is flawed. There is nothing inherently undesirable about resource-based industries provided they support high levels of productivity and productivity growth. Such industries can make a country wealthy if its resource position is highly favorable, as has been the case for Canada during most of its history. If resource-based industries continually upgrade their sophistication through improvements in their products and processes, competitive positions can be sustained and productivity growth insured* (Porter, 1991).

Productivity in the British Columbia forest sector is squeezed between a rising floor of raw material costs and a fixed ceiling for product prices set by international competitors in forest products industry and by the cost of substitute products. As a result, the British Columbia forest sector faces enormous challenges in maintaining the high levels of productivity which have produced the high standard of living enjoyed by the province,[7] let alone the high levels of productivity growth needed to sustain this standard of living in the face of the predictable, negative changes pressing the sector.

Historical reliance on old-growth timber, which required no human input to grow, meant the value of timber—the rent—was available for distribution to whatever parties found political favour. Historically, some of the rent flowed to licensees as an inducement to establish processing facilities.[8] Some flowed to labour in the form of higher wage rates. Figure 2.1 shows average labour costs for sawmills in the major North American producing regions. Despite almost a 20 percent devaluation of the Canadian dollar relative to the U.S. dollar over the last four years, B.C. Coastal producers (BCC) still support labour costs that are much higher than their competitors across the line (PNWW). Devaluation of the Canadian dollar combined with continued massive investments in labour-saving capital have worked to keep labour costs in the B.C. Interior (BCI) in line with the costs of their competitors in the U.S. South (USS).

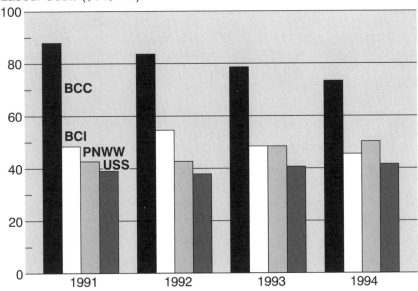

Figure 2.1   Labour costs ($US/mbf) in four North American regions, 1991-94

To counteract the negative competitive effects associated with rising timber costs and high wage rates, British Columbia needs higher-than-average productivity growth. Productivity growth can occur through reduction in the total costs of inputs or through increases in the value of outputs. As we have seen, there is endogenous upward pressure on raw material costs, and reductions in our high wage rates are not socially attractive. As a consequence, the only acceptable, sustainable answers seem to lie in more efficient production (Binkley, 1993).

Increases in productivity result from technology investments that are either earlier or better than those made by competitors. Yet in British Columbia the forest sector spends only about 0.7 percent of gross receipts on research and development where, for example, Sweden spends about 1.8 percent (Binkley and Watts, 1992).

Despite these problems, some elements of the forest sector have been able to adopt technology rapidly. Figure 2.2 shows the lumber recovery factor for BCI sawmills between 1986 and 1992. On average, this sector increased technical efficiency by about 1.4 percent annually, a remarkable performance by any standard. However, it is not clear that this increase in technical efficiency was adequate to offset higher production costs, competitor responses, and market-place effects (Binkley, 1994). Figure 2.3 shows that profit margins were no better than constant over the same period of these productivity increases.

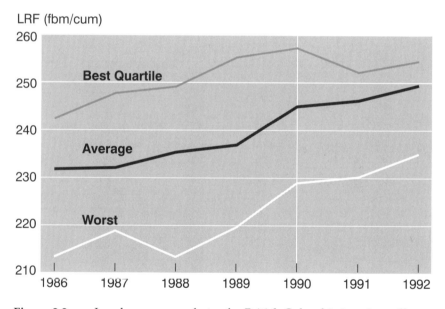

Figure 2.2   Lumber recovery factor for British Columbia interior mills, 1986-92

Sustaining Environmental Quality

Imagining that British Columbia devotes some 20 million hectares to zones where timber production is a significant or dominant use, then a vast area remains to support other forest uses. Urban and agricultural lands comprise a fairly small portion of the province, so perhaps 65 million hectares could be

Value Added ($1986/hr)   Productivity (cubic m/hr)

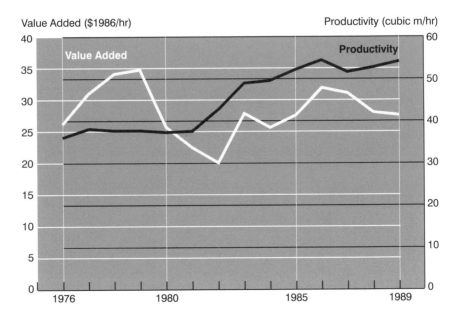

Figure 2.3   Value-added and productivity changes in British Columbia
                  sawmills, 1976-89

devoted, more or less exclusively, to sustaining the environmental values that
forests provide. This area would include at least 23 million hectares of productive forest, 20 million hectares of savannah and other vegetated land, and 20
million hectares of "rocks and ice". As a point of comparison, the area of France
equals about 55 million hectares. In other words, through a policy of zoning
and intensive, dominant-use management, British Columbia could devote an
area larger than France to sustaining non-timber values of the forest.

This policy might involve three zones: intensive timber production areas
as described above, strictly protected areas, and integrated management joint-
use areas to form the transition between the other two landscapes. Making
such a strategy work requires first that a core of protected areas is selected in a
way that includes reasonable representation of the great variety of British Columbia ecosystems in large enough blocks to sustain landscape-scale processes.
The current provincial Protected Areas Strategy is well suited to this task.

Second, the joint-use areas must be managed in a way that responds to
environmental needs. This nominally will occur through the Forest Practice
Code, but there may be serious environmental problems associated with the
approaches it specifies. The Code will reduce the average size of clearcuts,
and will require that areas logged "green up" (i.e., reach a specified age or
cover condition) before adjacent areas may be logged. While sounding in-
nocuous—if not beneficial—the net effect of these two provisions is to scatter

the harvest across the landscape. This pattern of harvests will fragment the forest with predictable consequences for biological diversity (Harris, 1984). The area of forest edge habitat will increase, and the area of forest interior habitat will decline. These landscape changes will tend to favour early successional species, while much of contemporary public concern focuses on late successional ones (e.g., the spotted owl).

The Forest Practice Code will also require more rapid development of the road system, and maintenance of a greater amount of it each year. The greater length of an active road network will provide more opportunity for illegal hunting. By building the road system faster, British Columbia forecloses the option of, in the future, deciding that an area should remain roadless. Since the very worst environmental problems associated with timber production in British Columbia come from the failure of roads, it is ironic that new legislation purporting to protect environmental values actually demands more road building.

Finally, the intensive management areas must be managed very intensively for timber production to make up the harvests lost through the creation of protected areas and low-intensity management areas. The Forest Practice Code logically should contain a very different set of standards for these intensive management areas than for the other zones. Control of the collateral environment damage frequently associated with logging should occur through land-use designations and not through management regulations on the intensive management areas. Such policy-driven constraints as minimum rotation ages, green-up/adjacency, and visual quality objectives should be relaxed in the intensive management areas. The tenure system should provide stronger incentives for investment in timber production on these areas. Applied research should be focused on these areas to adapt the lessons on intensive timber management from such locations as Brazil, New Zealand, the United States, and Sweden to the specific circumstances found in British Columbia.

NEEDED POLICY ADJUSTMENTS

Despite the vast size of the province and its relatively small population, land has become scarce in British Columbia. Each hectare seems to face multiple demands—from local residents, from the province as a whole, and from the international community. The increased scarcity of land logically implies that other productive factors—knowledge, capital, and labour—should be substituted for land to produce desired outcomes, whether environmental values or timber production. Government ownership and control of land in British Columbia means that the price system has not been able to signal the needed changes. Like a fault line that has accumulated strain over years of tectonic action, the British Columbia forest sector is close to rupturing. Abrupt change appears to be inevitable. Good policy would help ease the transition by providing flexibility for change. What are some appropriate actions?

Reform of Forest Institutions

A thin membrane of institutional arrangements mediates the interactions between humans and forests. In British Columbia, the most significant institution affecting forests is the tenure system. The basic concepts used in the current system of forest tenures are derived from the 1945 Royal Commission (the Sloan Commission). At the time, the basic societal need was to use old-growth forests as a wooden magnet to attract capital investment in the processing sector. Capital investment meant economic development, particularly in the hinterland of the province. The environmental values of the forest were not of great concern, in part because so vast an area of forests remained undeveloped.

Times have changed. Societal needs now involve attracting capital to invest in the forests themselves, in activities that produce value-added products, and activities that use such secondary sources of fibre as bark, sawdust, planer shavings and trim blocks. Standing timber is not a particularly logical or effective inducement for these kinds of investments, so the current tenure system is a blunt tool for crafting the future. As a result of increases in the timber fees collected by the provincial government and increasingly restrictive operating rules, forest tenures no longer are the potent plums of political advantage they once were.

The recent creation of Forest Renewal B.C. (FRBC) nominally provides the capital needed for investments in forest management. Unlike the many previous so-called "permanent" silviculture funds, increased stumpage fees and royalty payments directly fund FRBC without reference to annual appropriations from general revenues. An independent Board of Directors will direct these funds to five areas: silviculture investments, industry diversification and value-added manufacturing, environmental restoration, strengthening communities, and training workers. Although there are other similar arrangements elsewhere (e.g., the K-V funds for national forest lands in the U.S.), the magnitude of the FRBC fund is unique. Planning for FRBC assumed that the bell-wether grade of lumber (Interior spruce-pine-fir—SPF) would trade at 350 dollars US/million board feet—at this price, the pool of available funds would be about 400 million dollars US/year.

While preferable to the previous situation where little was invested in growing trees, the FRBC approach suffers four serious shortcomings. First, it is unlikely that the anticipated level of revenues will be realized. SPF has traded at around 300 dollars US/million board feet for the last three months, and respected industry analysts forecast even lower prices for 1995. At price levels below 300 dollars US/million board feet, the size of the fund is quite sensitive to fluctuations in product prices: each one dollar reduction in product price results in a 58 cent product-equivalent reduction in the stumpage rate (Binkley and Zhang, 1995). Falling product prices will jeopardize the entire program, and it is not at all clear that investments in the land will be able to secure priority allocation of whatever funds are available. Second, the Board is not constrained to invest in the best silvicultural opportunities; indeed, section 6(b) of the B.C. Forest Renewal Act *requires* the Board to "provide advice to Forest Renewal B.C. as to appropriate *regional goals* [emphasis added] for expenditures". The Board is contemplating devolution of funding authority to regional bodies. In the absence of clear guidance from the Act, the regional allocations are more apt to be governed by politics than by investment efficiency. Third, FRBC has no particular incentive to be frugal in how it administers its funds. For example, an early submission to the Board from the Ministry of Forests and the Ministry of Environment, Lands and Parks requested 13 million dollars per year to add 160 full-time equivalents to the two bureaucracies (Zak, personal communication). Finally, the efficacy of silvicultural investments depends on an intimate knowledge of the land base, local ecological conditions and overall landscape management objectives. It is doubtful that a distant, third-party funding agency will have adequate understanding of these local circumstance to invest wisely.

Discussions in British Columbia about forest tenures frequently use the analogy of owning a house outright (private land) versus renting it (tree farm licences or forest licences): if you lease a house to someone, you hardly expect them to paint the house (to invest in silviculture), and if you require them to do so (free-to-grow requirements) they will probably use the cheapest paint possible. To extend this analogy, FRBC offers to buy the paint but provides no incentive to apply it carefully or well (to implement the most efficient silvicultural investments).

At the same time, there is ample empirical evidence—both from elsewhere in the world, and now from British Columbia—that strengthened property rights lead to higher levels of private investment in forest management.[9] In British Columbia, stronger property rights could be achieved either through outright privatization of those lands in the intensive management zones created by CORE, or through the sale of long-term leases along the lines of New Zealand's recent policy. Existing tenure holders might retain their current lands, or might be satisfied simply to buy logs from others. New kinds of organizations would probably enter the field. For example, pension funds now own timberland in the United States, New Zealand, and Chile worth more than 4 billion dollars. The largest of the timberland management organizations serving pensions funds —the Hancock Timber Resource Group—now owns and manages over 40,000 hectares in British Columbia. Stronger private property rights would not only increase the pool of capital available to invest in forests themselves, but would also provide greater flexibility in responding to the rapidly changing circumstances now found in British Columbia.

Tenure reform should be accompanied by reform of other dysfunctional policies related to the tenure system. Perhaps the most significant of these are the appurtenant clauses in many coastal tenures that, on penalty of licence forfeiture, constrain logs from a specific licence to flow to a specific mill. The original objective of these clauses was to use timber to support the development of individual communities. But they have become very expensive subsidies for regional development. In times of tight log markets, efficient mills not associated with the restricted licences must close while the less efficient ones which are appurtenant to a particular licence continue to operate. The difference in log values between the efficient and inefficient mills is significant—in one case it amounts to about 60 dollars/cu.m. (Binkley, personal observation). In this case, the subsidy for each employee of the appurtenant mill is about 75,000 dollars per year, in addition to their wages. The author suspects that the employees involved would be pleased to receive a fraction of the capitalized value of this amount in return for agreeing to leave their jobs without complaint to the government.

Strengthening Transition Mechanisms

The policy changes required to put British Columbia's forest sector onto a path of economic, social, and environmental sustainability will inevitably produce losers as well as winners. Insofar as those old policies protect the self interest of potential losers, it is natural and predictable for them to hold fast to inappropriate policies and to slow the needed adjustments. The magnitude of the adjustments needed in the forest sector demand that explicit attention be paid to the problems of policy transition (Behn, 1978).

If policy changes improve economic efficiency (and many in British Columbia would), then—by definition—the winners would be able to compensate the losers and themselves still be better off than they were before the policy change. As a practical matter, those who wish to see policy changes—and will benefit from them —should support compensation to those who lose as a result of policy changes.

From a more philosophic vantage point, British Columbia has explicitly chosen *not* to allocate timber resources through market mechanisms, but rather through a process of governmental control. Timber prices do not properly signal the relative scarcity of timber, and firms' harvesting decisions are virtually dictated by government representing society as a whole. Firms deploy capital and labour moves to specific regions as a result of these socially determined decisions. Therefore, society as a whole logically bears the burden of unanticipated and unannounced adjustments to the system.

Compensation should be paid to those who are economically injured by reductions in timber harvests, changes in tenure policies, or other changes in forest policy which produce both winners and losers. Potentially injured parties include those who own capital or labour made obsolete by the policy change. Compensation should cover, for example, the reduced value of homes, the costs of necessary job retraining, the lost value of tenures, and the foregone returns to productive capital such as sawmills and pulpmills made redundant by the policy change. Compensation for lost tenures might involve granting stronger property rights to a licensee over a smaller portion of a current licence. Compensation is best paid in kind, so timber somewhere else (perhaps created by increased silvicultural investments) compensates for timber lost in one location, or another kind of job of comparable worth compensates for a job lost in logging or sawmilling.

Creating a Knowledge-Based Forest Sector

Rapid adoption of improved technology is key both to international competitiveness of British Columbia's forest sector, and to responsible stewardship of the environment (Binkley, 1993). Forest-sector research and development expenditures in British Columbia are small. A significant gap with our competitors exists both for forest-related and forest-products research and development (Binkley and Watts, 1992). Yet rapid development and adoption of leading-edge technology is a fundamental element of a sustainable future.

British Columbia firms face three particularly daunting tasks in effectively deploying research and development (Binkley, 1994). First, commodity-grade products (softwood construction lumber, market pulp, newsprint) dominate the British Columbia industry. Research and development leverage is less for commodities than it is for higher value-added products. By definition, commodities compete on the basis of production costs. Cost-reducing technology is likely to be available to all producers, so it creates no unique competitive advantage once adopted by all producers. As a result, only a strategy of rapid adoption will produce competitive advantage. Such a strategy is extremely difficult to implement. In this context, research and development may be better viewed as a powerful means of transforming the strategic direction of a company out of commodity businesses. MacMillan Bloedel's creation of Parallam® and Space Kraft®, and their Nexgen coated paper project are all examples of using research and development to move to value-added products.

Second, British Columbia's is a leading international exporter with a large share of many of the markets it serves. This market position exacerbates the problems of being a commodity producer. In these circumstances, some of the benefits of cost-saving technology are simply passed on to consumers. As long as the consumers are in one's own country, there is a case for governmental support of research and development, but when the customers reside elsewhere, such an argument obviously does not apply. In these circumstances, effective technology strategies again involve rapid adoption, and exploitation of features unique to British Columbia (e.g., western red cedar).

Third, as a result of the tenure system, British Columbia firms cannot exploit synergies between the design of forest management regimes and the design of new products and processes. Through its silvicultural regulations, the B.C. Ministry of Forests specifies the characteristics of future timber supply for the entire province—species mix, genotype, diameter, clear length—and all producers must adapt to these constraints regardless of their own market information. Under current institutional arrangements, the kind of research and development that led to the clearwood regime for *Pinus radiata* in New Zealand—an approach that permits fast-grown second growth timber to substitute in many uses for British Columbia's old growth *Pinus ponderosa*—would not be possible. Similarly, exceedingly low wood costs for Brazil's Aracruz pulp mill came from heavily targeted research and development to produce, plant, and process *Eucalyptus* spp. clones of high cellulose content. Such a strategy is unavailable to British Columbia firms operating on public lands.

A high-technology strategy for the forest sector will create benefits beyond those associated with the forest sector alone. The sector has some strong backward linkages with the high-technology sector which could be strengthened and exploited more effectively. Vancouver is the heart of an important international forestry services industry. The lower mainland is a hotbed of technology firms providing log- and lumber-scanning equipment and real-time sawmill optimization software. One equipment producer on Vancouver Island is among the world's leaders in the design and manufacture of cable logging equipment. Yet the provincial and federal governments have not, as a matter of policy, worked to build on these strengths.

Finally, a technology-based strategy will pay significant benefits for environmental quality. Better manufacturing efficiency is a powerful lever for environmental improvement. For example, from the perspective of product markets, just one year of the technological improvement depicted in Figure 2.2 is fully adequate to offset the reductions in timber harvests associated with the decision to set aside the Kitlope—a 320,000 hectare drainage on the North Coast thought to be the largest unlogged temperate rain forest watershed remaining anywhere in the world. As another example, progressively more sophisticated use of wood wastes has permitted substantial improvements in air quality in towns where sawmills operate. In the mid-1950s all slabs, edgings, planer shavings and sawdust (which collectively comprise about half of any log) for a typical Interior sawmill were burned as wastes. Then pulpmills operating on wood residues were established, leaving only the sawdust and planer shavings to be burned. Now technology has advanced further to permit the production of medium-density fibreboard—a high-quality panel used in furniture and similar end uses—from these residues. Removing all of the wood fibre from the waste stream will leave only the bark, and mills will burn much of this for process heat used in the manufacture of lumber, MDF or pulp.

Improved silvicultural technology—from better inventory and yield information to sophisticated techniques of molecular genetics—can sustain harvest levels on a smaller land base, freeing land for allocation to other uses. The power of this technology has not been extensively used in British Columbia, but has in other parts of the world. For example, because of an aggressive, high-technology plantation program, forest companies in New Zealand no longer log in that country's native forests, but instead rely entirely on plantation forests. Their agreement to refrain from logging in natural forests—the Tasman Forest Accord—had virtually no economic impact on the country. In contrast, such an agreement in British Columbia would close over 90 percent of the forest industry, largely because the province has made no similar investments in research and development and forest management.

CONCLUSIONS

In husbanding its forests, British Columbia faces an ancient challenge. The paths traveled by earlier societies are known, but their destinations have often been unpleasant. Past policy and development in British Columbia have brought the province to a cross road in the forest. Only "the one less traveled by" will create a sustainable forest-based economy for the future while maintaining the critical, renowned ecological and environmental features of British Columbia's magnificent forested landscape.

Following this path will require a massive redirection of current policies, both public and private. Many policies that have served well in the past are dysfunctional guides to the future.

Some of the needed changes are now under way. Land-use planning through the CORE process will provide greater long-term political certainty in the forest sector. Increased certainty is prerequisite to the high level of capital investment—in forests and in new, sophisticated processing equipment that sustainability, in its broadest sense, requires. The Forest Practice Code will provide a framework for guiding management in the different land-use zones. FRBC may be able to provide the capital required to finance this transition.

But these policy changes must be carefully implemented and strongly reenforced if they are to be successful. Once land-use zones have been established, various interests will no doubt seek to poach across the boundaries. Government must wisely distinguish legitimate needs to revise land-use zones from simple rent-seeking. Economic instruments such as those increasingly used for pollution abatement may be helpful in drawing these distinctions.

The management rules for the various zones must be carefully crafted to support the distinct management objectives of each zone. Just as industrial intrusion on protected areas should be strictly limited, regulatory intrusion on intensive management areas should be carefully proscribed. Differences in Forest Practice Code regulations among the various zones provide a useful measure of success in this regard.

Once land-use zones have been defined and zone-specific Forest Practice Code rules have been written, it will be possible to craft a set of institutional arrangements and land tenures which more productively serve the sector. These arrangements logically involve significant public control in zones where public values dominate, and significant private control in zones where private values are most significant. In the case of parks and protected areas, institutional reform will require a much strengthened parks agency to handle the capital investment and management activity needed to insure that the protected areas and very low intensity zones do, in fact, provide the environmental values anticipated from them. In the case of integrated-resource management zones, institutional reform will require either strengthened direct public management activities, or a much more sophisticated set of licence documents than are currently used. Economic instruments which bring licensee and public interests into line merit careful attention (e.g., pricing systems for recreation, water flows, or site degradation). In the case of intensive timber management areas, tenure reform will require more powerful inducements to make highly effective investments in timber production. Ample evidence from British Columbia and elsewhere in the world suggests that markets operating through ordinary private property rights provide the needed incentives. Experience from elsewhere in the world—especially New Zealand, Sweden and the United States —will provide useful guidance on the advantages of different kinds of private ownership and of different mixes among the kinds of private ownership (small and large; institutional, industrial and individual), but the unique circumstances in British Columbia will no doubt require unique approaches to strengthening private property rights in forest land.

Change is always uncomfortable, and when the stakes are as high as they are in the British Columbia forest sector, discomfort invites paralysis. Sensitive attention to the problems of transition can reduce the discomfort and invite more rapid, creative, and positive responses from the parties involved. Compensation principles should be articulated at the outset. New tenure arrangements should be made sufficiently attractive so at least some licensees will voluntarily adopt them. Because no one really knows the optimal approach for a sustainable future, experimentation (and its concomitant, failure) should be encouraged. Different approaches might suit different areas.

To make progress down the new road—the one that leads towards a sustainable future—will mostly require new ways of thinking about old problems. The forest industry must embrace the righteousness of forest conservation and preservation, and environmentalists must accept the desirability of a robust, efficient forest products industry. Governments must respect the stewardship capacity of the private sector, and the private sector must respect the necessity of governmental regulation of the public goods produced by forests. This revolution of the mind will not be easy, but is the only means down the path less traveled by.

ENDNOTES

1   In broad lines this is consistent with the development of the forest sector in the United States (Clawson, 1979; Sedjo, 1990). Harvest exceeded growth until the 1950s. Early in this century, the trends were such that the famous American conservationist Gifford Pinchot (1907: 3) predicted: "If we accept the larger estimate of annual growth and apply it to the present rate of consumption, the result shows a probable duration of our supplies of lumber of not more than 33 years . . . [It] is certain that the United States has already crossed the verge of a timber famine." Increased scarcity drove up timber prices, and high prices forestalled the predicted timber famine by choking off demand for wood products and encouraging investments in forest management. Timber prices rose at a real rate of about 4.6 percent/year between 1910 and World War II and by about 3.1 percent from that period to the mid-1980s (Binkley and Vincent, 1988).

2   This assumes that the cost of harvesting the last old growth is greater than its value once harvested. See Clark (1973) or Page (1977) for a discussion of the economics of extinction.

3   Evidently, the relationship between income and environmental values is not new. Perlin (1991: 120) comments:

   Seneca best articulated the romantic view of forests shared by many of the leisure class of his time: "If you ever have come upon a grove that is full of ancient trees which have grown to unusual height, shutting out the view of the sky by a veil of pleated and intertwining branches, then the loftiness of the forest, the seclusion of the spot and the thick, unbroken shade on the midst of open space will prove to you the presence of God."

   One cannot help but note the similarity between this comment, made nearly two millennia ago, and contemporary descriptions of old-growth forests in British Columbia.

4   My own analysis of 22 of the 36 reports suggests a 20-year reduction in AAC in the Interior of 11.8 percent and on the Coast of 25.1 percent, for a reduction in the provincial total AAC of 15.9 percent.

5   Park (1991) estimated the economic impact of the British Columbia forest industry in the metropolitan area of Vancouver for 1989 to be 6.0 billion dollars of GDP and 115,000 jobs with wages and salaries of 3.0 billion dollars. A more recent study (Chancellor Partners, 1994) found that in 1993 133,000 jobs (one in six) and 6.2 billion dollars of regional GDP in metropolitan Vancouver depended on the forest sector.

6   This study of the Revelstoke Forest District found that, through moderately increased management intensity, about 40 percent of the land base would produce the same amount of timber as would 100 percent of the land base under current rules for integrated resource management.

7   At present, mills in British Columbia are, on average, mediocre competitors in newsprint and pulp but, at least in the Interior, are international competitors in lumber (NLK, 1992; Simons, 1992).

8   Before the April 1994 changes in the stumpage rates, analysts commonly estimated the uncollected rents at about 10 dollars per cubic metre. The increased stumpage payments announced in April averaged about 11 dollars per cubic metre, so the amount of uncollected rent, especially for licencees with marginal mills, is probably now small. This is roughly consistent with the estimates of Binkley and Zhang (1995), which found that the change in the stumpage system reduced the capital value of publicly traded firms in British Columbia by about 1.3 billion dollars (or about 2.8 billion dollars if grossed up to the sector as a whole).

9   A recent study (Zhang, 1994) empirically confirmed this well-known theoretical argument. Zhang examined expenditures on silvicultural activities on different kinds of tenures while holding a variety of factors that might affect such investments (e.g., biogeoclimatic zone, site quality, location) constant. After controlling for all of these factors, silvicultural.

REFERENCES

Alverson, W.S., Kuhlman, W., and Wallen, D.M. (1994). *Wild Forests: Conservation Biology and Public Policy*. Washington, DC: Island Press.

Behn, R.D. (1978). How to terminate a public policy: A dozen hints for a would-be terminator. *Policy Analysis*, Summer, 393-413.

Binkley, C.S. (1994). Designing an effective forest sector research strategy for Canada. Unpublished manuscript, Faculty of Forestry, University of British Columbia, Vancouver, B.C.

Binkley, C.S., and Zhang, D. (1994). The impact of timber-fee increases on B.C. forest products companies. Unpublished manuscript, Faculty of Forestry, University of British Columbia, Vancouver, B.C.

Binkley, C.S., Percy, M., Thompson, W.A., and Vertinsky, I.B. (1994). A general equilibrium analysis of the economic impact of a reduction in harvest levels in British Columbia. *Forestry Chronicles*, 70, 449-454.

Binkley, C.S., and Vincent, J.R. (1988). Timber prices in the US South: Past trends and outlook for the future. *Scientific Journal of Applied Forestry*, 12, 15-18.

Binkley, C.S., and Watts, S.B. (1992). The status of forestry research in British Columbia. *Forestry Chronicles*, 68, 730-735.

Binkley, C.S. (1993). Creating a knowledge-based forest sector. *Forestry Chronicles*, 69, 294-299.

Chancellor Partners. (1994). The economic impact of the forest industry on metropolitan Vancouver. Prepared for the Vancouver Board of Trade, Vancouver, B.C.

Clark, C.W. (1973). Profit maximization and the extinction of animal species. *Journal of Political Economics*, 81, 950-961.

Clawson, M. (1979). Forests in the long sweep of American history. *Science*, 204, 1168-1174.

Farnum, P., Timmis, R., and Kulp, J.L. (1983). The biotechnology of forest yield. *Science*, 219, 694-702.

Harris, L.D. (1984). *The fragmented forest*. Chicago: University of Chicago Press.

Lyon, K.S. (1981). Mining of the forest and the time path of the price of timber. *Journal of Environmental Economics and Management*, 8, 330-344.

McNeely, J.A. (1993). Lessons from the past: Forests and biodiversity. Unpublished manuscript, IUCN, Gland, Switzerland.

Ministry of Finance and Corporate Relations. (1992). British Columbia community dependencies. Prepared by the Planning and Statistics Division for the Forest Resources Commission.

Nixon, R. (1991). Comparative data charts explain forest management policies. *Forest Planning Canada*, 7, 32-45.

NLK (1992). *The pulp and paper sector in British Columbia*. Discussion paper prepared for the Forest Summit Conference, Vancouver, B.C.

Page, T. (1977). *Conservation and economic efficiency.* Baltimore, MD: Johns Hopkins University Press.

Park, D.E. (1991). The forest industry's role in Vancouver's economy. Vancouver Board of Trade, Vancouver, B.C.

Perlin, J. (1991). *A forest journey: The role of wood in the development of civilization.* Cambridge, MA: Harvard University Press.

Pinchot, G. (1907). Conservation of natural resources. *The Outlook,* 12 October, 1907.

Porter, M. (1991). Canada at the crossroads: The reality of a new competitive environment. A study prepared for the Business Council on National Issues and Government of Canada. Harvard Business School and Monitor Company.

Porter, M. (1990). *The competitive advantage of nations.* New York: The Free Press.

Sahajananthan, S. (1994). *Single and multiple use of forest lands in British Columbia: The case of the Revelstoke Forest District.* Report submitted to B.C. Ministry of Forests, Revelstoke Forest District, March.

Sedjo, R.A., and Lyon, K.S. (1990). *The long-term adequacy of world timber supply.* Washington, DC: Johns Hopkins University Press.

Sedjo, R. (1990). The nation's forest resources. *Discussion Paper* ENR 90-07, Resources for the Future Inc., Washington, DC.

Simons, H.A. (1992). *The wood products sector in British Columbia.* Discussion paper prepared for the Forest Summit Conference, Vancouver, B.C.

Swallow, S.K., Parks, P.J., and Wear, D.N. (1990). Policy relevant nonconvexities in the production of multiple forest benefits. *Journal of Environmental Economics and Management,* 19, 264-280.

Tolnai, S. (1991). Addition value to our heritage through silviculture. Paper presented to Western Silvicultural Contractors Association, 5 February 1991, Vancouver, B.C.

Vincent, J.R., and Binkley, C.S. (1993). Efficient multiple-use forestry may require land-use specialization. *Land Economics,* 69, 370-376.

Vincent, J.R. and Binkley, C.S. (1992). Forest-based industrialization: A dynamic perspective. In N.P. Sharma (Ed.), *Managing the world's forests* (Chapter 6). Dubuque, IA: Kendall/Hunt, for the World Bank.

Zhang, D. (1994). *Implications of tenure for forest land value and management in British Columbia.* Unpublished Ph.D. thesis, Faculty of Forestry, University of British Columbia, Vancouver, B.C.

Plate 4 (overleaf)   A Douglas Fir about to be felled   ➤

Reinventing British Columbia: Towards a New Political Economy in the Forest

R. Michael M'Gonigle[1]

Faculty of Law, University of Victoria

Throughout British Columbia, conflicts have long raged over the future of the province's forest lands. Some of these conflicts—from Carmannah to Clayoquot Sound—have been high profile, international issues. Many others have been largely out of public sight and sound—a little dispute over a cutting permit in a community watershed, an ongoing debate between a small northern Native band and the Ministry of Forests over logging, or preserving a landscape for traditional uses. In this context, the pioneering work of Harold Innis still has much to teach us. Certainly we can learn from his analysis of the dynamics of our economic dependence on such "staples" as fish, fur and timber (Innis, 1956), but we should also remember his larger concern to understand more broadly the long historic patterns of power in the West (Innis, 1950). Most important today, however, is to understand how his analysis might be updated to take fuller account of the dynamics of global resource exhaustion and ecological decline. This is especially important insofar as the field of Canadian political economy has really failed to take ecology seriously, instead attempting to "add on" environmental issues as just another subset in a largely unchanged analytical frame (Williams, 1992). In this chapter, an effort is made to shift this frame within the context of British Columbia forest politics, a politics which reflects not just the economic hegemony of the Innisian core but the unsustainability of centralist power itself.

ECOLOGICAL CENTRALISM

The starting premise of this analysis is simple: we are all here participating in a social and economic structure that is compulsively and fundamentally unsustainable, and has been for some time. That structure has now attained global dimensions; space to absorb the consequences of its unsustainable activities is running out. As a result, that structure requires not just incremental reform, but structural re-formation. As written recently:

> *Looking back in history, we recognize that every age has, under*
> *changing circumstances, been faced with the unexpected, unwel-*
> *come challenge of radically adjusting the way it does things. In*
> *school, we all learn about the Scientific and Industrial Revolutions*
> *and, standing on the afterside of this history, we understand why*
> *these revolutions occurred. Change is inevitable.*
>
> *Today, the dominant fact of global life is that we are catastrophical-*
> *ly overshooting our resource base. We take what nature provides,*
> *often at a dizzying pace, seemingly oblivious to the vastly different*
> *environment that awaits us . . . . If our wholesale plunder of old*
> *growth and the social dislocation that flows from it isn't what we*
> *want, then we must change the way we live. We must embark on*
> *an Ecological Revolution, a revolution equal in scale to the great*
> *revolutions of the past, only this time we stand not after the revolu-*
> *tion, but before it* (M'Gonigle and Parfitt, 1994: 14-15).

Revolutions, by definition, involve re-formation of an inherited structure of power, but the Scientific and Industrial Revolutions mentioned above are not revolutions in the traditional political sense, of peasants storming the palace. They are larger than that, whole social and cultural changes affecting our basic intellectual understanding and our deep-seated cultural values, as well as our political and economic processes. The revolution that awaits us in the British Columbia forests, and worldwide, is a revolution against what might be called "ecological centralism." And, like the great social transformations of the past, it offers huge opportunities. But these opportunities are now largely ignored, and very little understood. They will become commonplace expectations only to those post-revolution scientists, industrialists, or ecologists who will follow us.

Here is a brief characterization of the pre-revolutionary structure which has now become so problematic. Its signal characteristic is its unsustainable *centralism*. This centralism is manifest in a variety of ways. Geographically, it is supported by Innis-style resource flows from the hinterland to the heartland, from the periphery to the core, from rural to urban. The resource rents accrue to the office towers in downtown Vancouver and Victoria, Toronto and Tokyo. Politically, it is dominated by an integrated hierarchical structure of power, whether this be corporate offices or Crown bureaucracies, where authority is concentrated at the top. Economically, it is characterized by flows of energy, materials, and wealth that are *linear*, moving from one place to accrue to a distant locale, down the freeway and into the cities. These linear flows displace more traditional processes that are *circular*, people exchanging with each other locally, their mutuality sustaining communities in place wherever they might be (Figure 3.1). (This difference explains, for example, our social obsession with creating physical wealth through resource-intensive *trade-based* economies of scale, rather than creating social and community wealth through *multiplier-based* economies that recirculate smaller volumes of materials and capital in place.) Socially and culturally, centralism is dominated by the top-down intelligence of technocratic expertise and management, rather than the bottom-local experience of living with, and in, community. In an updated, ecological version of the core-periphery analysis, this social structure of power might be conceptualized as a structure of Central, rather than Territorial, power.

In many ways, this characterization bears similarities to the long-standing concerns of staples theorists, though there are considerable differences. What is proposed here builds on, but is more than a modification of, such theory. For one thing, the staples theory which posits a distinction between "core" and "periphery" implicitly marginalizes the latter by that very terminology, or similar terminology such as "hinterland". In contrast, opposing Territory against Centre situates our political economy in a fuller dialectic that doesn't privilege one over the other. The concepts also resonate broadly, for example, in forming a range of conceptions from Crown sovereignty (a quintessentially Centrist invention) to the "alternatives movement" (and its distinctly Territorial flavour). Indeed, Centre and Territory are not so much geographical ideas as they are forms and tendencies of social organization. Territory definitely has a physical component, bred of its character as physically and community-rooted experience. But elements of Territory are manifest everywhere (in close-knit neighbourhoods in the city, for example) as are Centrist forms (such as the satellite dish behind the remote Prairie farm house). Thus the structures intermingle, as do the dynamics. Of critical importance in making our future is maintaining the balance or, more accurately, redressing the imbalance, between them.

Centrist Model   **Territorial Model**

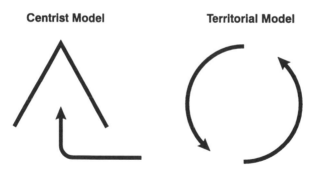

Figure 3.1   Basic character of linear vs circular energy flows

Today, Centrist structures dominate our lives wherever we live, and they pervasively shape our physical experiences and intellectual ambitions. Elite academics and intellectuals talk to each other and publish in refereed journals, largely unaware that, in so doing, they seek to interpret between themselves a desktop world increasingly remote from a world of direct physical experience. Senior corporate executives draft strategic plans for their companies, compulsively reciting a mantra to the only "natural laws" that count—competitive productivity and economic growth—heedless of the real-world social and environmental costs of these laws. Labour leaders attach the futures of their unions to this same exploitative treadmill—forgetful of their own social history —in their quest to safeguard their institutional positions against more basic changes in the economic landscape. Government bureaucrats are compelled to manage rationally, to control the world from the top, cautious to avoid the unsettling consequences that would accompany any challenge to our hierarchical traditions.

This structure of central power is, above all, therefore, self-maintaining. In this, it is an ecological (or *anti-ecological*) centralism sustained by distant flows of energy and materials out of local environments and communities, and at great cost to them. As Rees and Wackernagel explain, our "ecological footprint" is far larger than the place where we live because we draw so many more resources from afar than we could produce locally—we "import sustainability" (Wackernagel and Rees, 1995). This has, indeed, long been the history of the colonial expansion that Innis analysed; it is, in fact, the history of the "rise and fall" of countless civilizations of the past (Frank and Gills, 1993). It is this pattern and process that characterizes the past—and present—of the British Columbia forest economy.

In British Columbia and elsewhere, anti-centralist challenges come from the social movements which are its victims. The 800 protesters arrested at Clayoquot Sound were vocally critical of the big powers, of corporate, bureaucratic and labour organization that they confronted, reminiscent in their language and their ardour of the labour protests of the nineteenth and early twentieth centuries. Similarly, the so-called international "boycott" campaigns of Greenpeace and Sierra Club that seek to make consumers of wood products in Europe aware of the costs being inflicted by their purchases, attempt to meet the long-distance power of the multinationals on their own terms by informing remote consumers of the full local costs of their consumption patterns. Similarly, First Nations who resist multinational logging in their traditional territories—from the Haida blockades on the Queen Charlotte Islands (Haida Gwaii) to the court cases of the Gitksan-Wet'suwet'en—seek to re-balance territorial against central power.

THE CURRENT FORESTRY DEBATE

This critique may seem harsh or extreme given the great changes which have been made under the New Democratic Government in British Columbia since 1991. First came the Commission on Resources and the Environment (CORE), a several years-long participatory planning process in three regions of the province (Vancouver Island on the coast, the Cariboo-Chilcotin region in the province's south central interior, and the Kootenays in the province's southeastern mountains.) In combination with the Ministry of Environment's Protected Areas Strategies, these processes have been aimed at identifying new park and wilderness areas for protection, up to a goal of 12 percent of the provincial land base by the year 2000.

Second has been an internal Ministry of Forests recalculation of the inventory of standing timber available for logging in the province, and a consequent reduction in the level of allowable cut to make up for past over-cutting. This has led to some serious reductions in cut levels in local districts. Third has been the creation of a new set of logging standards through the Forest Practices Code, which promises to incorporate environmental constraints into everyday logging practices, leading to a further reduction in available timber, and an increase in costs associated with logging. Fourth has been the inauguration of a Forest Renewal Plan, financed by an increase in stumpage rates, that will direct several hundred millions of dollars annually into enhanced silviculture, community development and new economic enterprises.

This is clearly an impressive list of governmental initiatives, and is widely regarded as the crowning achievement of the NDP's term of office. Nevertheless, taken together, these initiatives amount neither to a structural attack on the problem of centralist unsustainability in the British Columbia forest economy, nor to a strategic initiative to mobilize public sentiment in favour of such action.

On the ground, the impacts of the changes are limited. To start, it must be appreciated that protecting 12 percent of the land base doesn't actually create anything new on the ground; we aren't actually creating wilderness. It exists there already, and in greater quantity than 12 percent of the land base. This number is just a fence—a minimum area—where the industrial machine is supposed to stop. Everyday, the amount of natural habitat continues to shrink at a dramatic rate, and there is, in any event, no certainty at all that the 12 percent ultimately "protected" in islands of preservation will actually be able to survive as pristine playpens of biodiversity surrounded by denuded landscapes. And where is this fence to be planted? Originally, that protected 12 percent of the land base was intended to be "representative" of a range of ecosystems. As it has turned out, however, valuable old growth stands of forest have been excluded and will be under-represented in the final wilderness mosaic (in the range of 68 percent) with a predominance of wilderness to be set aside in the higher elevations of "rock and ice" where the forest industry has no interest. Meanwhile, as one set of cut levels decreases due to the timber supply review, another set of cutting rights is waiting to go up with the potential activation of long-neglected "pulpwood agreements", and the push of the industry into new areas and forest types.[2]

Outside the wilderness areas, logging regimes of different intensities are to be implemented after the CORE process, but implemented by an as-yet-unclear administrative process. Existing land tenures controlled by the forest majors—the predominant form of corporate control and power—are not "on the table", and will not be. Instead, eco-tourism operators hoping to gain access to Crown lands for new operations confront a glacial bureaucratic pace in developing new policies that might shift lands, and community economies, to non-forestry uses.[3] Similarly, communities anxious to protect their community watersheds as a source of clean drinking water find themselves without legislative protection as logging companies press to get access to the timber resources still remaining in these sensitive areas.[4]

In addition to the marginal shift on the ground is a marginal shift in the corridors of power. The consensus-based CORE process was hailed as a new way of making decisions outside the old closed-door political forum. Backed with neither sufficient financial resources nor real political authority to empower new groups, however, the process encountered a forest industry (both corporate and labour) that largely bided its time in the negotiations, blocked consensus, and then waged an old-style political campaign to influence Cabinet in its decision-making after the CORE reports came out. With its three regional processes now complete, CORE proposes to institutionalize its decision-making process more formally through a *Sustainability Act*, a potentially promising (if largely process-oriented) initiative, but one that is not on the legislative agenda. At the same time, in the review and approval of development plans and cutting permits, the new Forest Practices Code continues to grant huge levels of discretionary decision-making authority to forest ministry District Managers, the same managers who have, in the past, often been closely aligned to the industry they were intended to regulate.[5] Despite the changes, the government remains concerned to maintain timber supplies to existing mills. It has even been alleged that it was expediting the approval of cutting permits in a way reminiscent of the much-criticized Social Credit policy of "sympathetic administration" (Palmer, 1995). These incremental changes only foretell greater frustration, and conflict, in the years ahead.

In contrast, little active development of alternative forms of development is taking place. The 1993 report of Select Standing Committee on Forests, Energy, Mines and Petroleum Resources, *Lumber Remanufacturing in British Columbia*, which recommended improved timber access for the province's small business, value-added sector, has been quietly shelved. At the time of its release, it was roundly condemned by both the Council of Forest Industries (COFI) and the International Woodworkers of America (IWA) and was then referred by the Forests Minister to an industry-dominated, in-house advisory committee.[6] Community forests, long a favourite alternative with environmentalists and "new foresters", are nowhere to be seen, even though the 1991 Forest Resources Commission recommended that existing corporate tenures be cut back by 50 percent, and replaced with a variety of new arrangements. Instead, 17 of the ubiquitous Tree Farm License tenures, which provide the industry giants with a secure contractual base of supply, have been "rolled over" into new 25-year terms without public hearings or public debate. This smooth updating of tenure is the price which the government has had to pay for the industry's acquiescence to other changes; it will help lock in, for another generation, the corporate control of the land-base for the final liquidation of the provincial forest.

A far-reaching initiative is, of course, the Forest Renewal Plan introduced in 1993. This plan contains the seeds of a good alternative economic development strategy, and it is well-funded from an increase in stumpage fees charged to the industry. The plan is, however, limited in its intent to the mitigation of the effects of wilderness set-asides, a vehicle to give substance to the Premier's promise that "not a single job will be lost" to wilderness set-asides. It is not a pro-active initiative that offers a new, community and small business-based, economic development strategy to replace existing corporate holdings. Quite the contrary, new initiatives will be approved by a provincial Board set up by Victoria, the resource rents to pay for the programs still flowing up and through the Centre, not recycling in the Territory. And, indeed, the best paper objectives notwithstanding, the early allocations of funding under the Plan have gone right back to the industry in the form of silvicultural investments and environmental restoration projects, with almost no money being allocated in the Plan's first year of operation to community diversification or small business value-added manufacture.

This analysis is not intended to denigrate these governmental initiatives, which constitute the broadest, most environmentally-oriented range of governmental reforms ever undertaken for the British Columbia forest industry. This may well be the NDP's prime legacy from their term in office. Nevertheless, our concern is *structural* reform and, here, the package simply does not measure up. In fact, in a world of rational policy-making (admittedly a fictional world), the procession of policies should have been quite different with land use planning *and tenure reform* going hand in glove. On that basis, a relevant process and set of forest practice standards and economic transition policies would be constructed *subsequently*, tailor-made to the new, not the old, structure of land control. Given the real world dynamics in the liberal/administrative state, however, the fundamental questions remained unasked, and unanswered. If so, perhaps the dominant lesson of the NDP's term in office will not be what it could accomplish, but what it couldn't. Is state-directed, structural ecological change even possible?

A STRATEGY FOR CHANGE

The answer to this question is "Yes": a strategy for structural change is possible. An economically efficient, and ecologically sustainable, alternative does exist. However, to achieve it, we must be willing to consider not just reform, but re-formation. In addition, we must look beyond the state to make it happen.

In other writings, it has been argued that we must begin this re-formation by rejecting the idea of "sustainable development", a concept where, as many environmentalists point out, "they get the noun, and we get the adjective". Development in the traditional mode is simply not compatible with the constraints of environmentalism. In its place, we must turn this concept around, and begin to "develop sustainability". In this formulation, "they get the verb, and we get the object". Thus, we might still have economic growth, but growth to a new non-growth state—transformative growth. The goal is to change the very dynamics of our politics and our economics so that they will inherently move us towards, not away from, sustainability.

Parallel with this rejection of sustainable development comes the need to recognize the broader source of our inertia—our dependence still on the liberal state. The promise of liberalism itself is dead, but we refuse to bury it. Philosophically, the liberal individualistic creed (that eschews either a concern for external moral justification, such as communal or environmental sustainability, or an attention to real world power relations) has been withering under a sustained theoretical attack for over a decade now.[7] Similarly, the liberal market ideology (rooted especially in neoclassical economic theory) has been thoroughly discredited as a coherent intellectual framework both from within the discipline, and from outside it, by ecological economic theorists (Knetsch, 1995; O'Connor, 1994). As importantly, we must address the failed role of the state in charting any sort of collectively agreed-upon course in the face of private corporate power, entrenched bureaucratic inertia, and the self-interested tussle of elite group competition.

As global corporatism triumphs, and sustainability withers, why do progressive critics (especially those in the Innisian mode) hesitate to move beyond the maladapted legacy of the liberal state, and its social democratic variant? Because, for one thing, we are all involved in it. Our experiences and analysis are shaped by its inevitably Centralist perspective. We can accept the passing of this world only when we are freely able to understand and embrace the possibility of something else. In short, we cannot reject what we know until first we can envision something to take its place; only then can we face the dysfunctions of a totalistic system from which we take our sustenance, our individual and collective identities, and our back-seat back-window view to the future.

REINVENTING BRITISH COLUMBIA

This chapter does not attempt to relate the details of the possible economic/ ecological transition, so much as to highlight the architecture of such a strategy. This discussion draws extensively on the analysis found in *Forestopia: A Practical Guide to the New Forest Economy*, where more specific details can be found.

In the age of "hidden subsidies" and international competition, any ecological political economy must begin with the recognition that the contemporary global market economy is, and has long been, underwritten by the broad subsidies of environmental and social decline. In British Columbia's forest economy, this takes a diversity of forms in, for example, the eradication of millennia-aged old growth forest ecosystems, the destruction of fisheries habitat and species, the emasculation of traditional Native cultures, the erosion of small non-Native communities, the continuing displacement of forestry workers, and so on. We speak of economic subsidy because the "true values" of this economic activity are not factored into the price of forest products. Indeed, this has been characterized as a "value-subtracted" economy because the final products are valued far below what would have been returned by a less controlled, less oligopolistic system. This structural trait contrasts sharply with the incremental demands today for "value-added" processing by otherwise unchanged and very wasteful, big-scale processes (M'Gonigle and Parfitt, 1994: Chapters 2-4). This understanding reflects clearly the insight of the ecological economic literature, which posits the need to reduce GDP by a value that reflects the resource depletion in order to offset the otherwise inflated financial figures from industries dependent on resource liquidation.[8]

In contrast to this subsidized economy, the British Columbia forest economy needs to move in the direction of an "alternative" economy. This is not an accidental term. Forestry "alternatives" are but one part of a holistic shift in our centralized structures, such a shift being best explicated by the so-called "alternatives movement" encompassing sectors as diverse as "alternative energy" and "alternative health". In the terminology of political economy (which looks at structures and flows of power) the essence of this shift is from *linear* to *circular* flows and institutions. This contrast is explained in *Forestopia* where a distinction is made between the existing "volume economy" of high resource throughput generating limited social welfare with a proposed "value economy" where resources are husbanded so as to extract maximum social benefits with the least ecological costs:

> Imagining healthy growth in practical terms is not a difficult exercise. As we have seen, the processes underlying the volume economy are extractive and linear—getting the wood out from one place to supply bigger mills someplace else. To sustain our social organization, which is addicted to this linear growth, the volume economy achieves its wealth by sapping communities and environments of

*theirs. The value economy, in contrast, assumes that ecological and
community processes are circular. That is, to be sustainable, these
processes must maintain themselves, living on the stock of natural
and social capital with which they have been endowed, so that they
can return long-term stability to the forest and long-term value to
the local community* (M'Gonigle and Parfitt, 1994: 53-54).

This analysis of resource flows clearly has much in common with traditional, core-periphery staples theory. Its specifically ecological character, however, is found in its strongly prescriptive message that economic development
doesn't just mean shifting to higher levels of manufacturing with these flows,
but reducing the level of flow itself. In short, the ecological prescription for
the staples crisis is that human society everywhere needs to move toward more
circular systems—ecologically, economically, even constitutionally. Just as forest
resource rents should be retained locally in British Columbia, so too urban
areas must learn to conserve and recycle the resources—newsprint, lumber,
energy, minerals, chemicals, water—which they have hitherto drawn heed-
lessly from the hinterland.

Forcing such structural changes in forestry (and other resource sectors) is
the larger promise of First Nations in their assertion of aboriginal title. To the
extent that Native self-government for, say, the James Bay Cree, can drive up
the price of energy for Montreal, then all the complex technocratic cost-benefit
analyses in planning Montreal's energy future will shift along the sustainability continuum. Where veto powers are geographically decentralized, the city
(and its downstream aluminium plants) will inevitably be forced to conserve
as a result of the inherently higher costs that are imposed by the geographical
realignment of decision-making power. This conservation is exactly the message that "alternative energy" advocates have been preaching for over two
decades. The same can be said for "alternative forestry" and community-based
fisheries. Instead, of course, the natural reaction to these initiatives is to see
First Nations not as catalysts for a larger transformation, but as threats to the
status quo, threats to be co-opted by drawing First Nations into the centralist
system of political decision-making (via delegated authority) and economic
development (via joint ventures) (M'Gonigle, 1992).

Moving from a linear to a circular economy in the British Columbia forest
economy has three specific components. First is the need explained above to
shift from a volume-based (i.e., high resource throughput) economy to a value-
based (i.e., low throughput) economy. Second is the need to move from capital-
intensive to labour-intensive modes of production. In British Columbia, this
dichotomy is most evident in the conflict between the high wage strategy of
the IWA (and its alliance with high throughput multinational corporations),
and the low throughput small business, "value-added" sector.[9] Third is the
need to shift from central (corporate and bureaucratic) to community-based
forms of management and control.

These shifts play out in many forms. In the woods, it means moving away from large corporate fibre farms to community forests and private or public woodlots. This involves a shift in technology as well, away from high-capital clearcutting to a reliance of more selective forms of (here is that word again) "alternative" technology. This shift is mirrored in the manufacturing sector as well where large, mass-production companies still block the institution-displacing growth of smaller, more nimble, value-added businesses. This shift is part of what more mainstream critics call the need to move from the Old to the New Economy, or from the Old to the New Competition (Best, 1991).

To allow these changes to take place requires, above all, a dismantling of the corporate tenure system which today allows large companies to control the cutting and use of vast areas of forest. Instead, these tenures should be replaced with a network of individual woodlots, community forests, and First Nations territories, the output of which would be sold through regional log markets, with the returns from sales accruing back to the many forest stewards, and the resource rents accruing to community resource boards in the regions. Such a market system is a radical suggestion, as it would undermine the constipated corporate and bureaucratic hegemony that today dominates the forest landscape in forestry, as in other sectors, and stifles both economic innovation and environmental/social stewardship. In conjunction with a community-based devolution of authority rooted in ecosystem-based management, this implies a dramatically new role for the state, a more efficient role as the facilitator and protector of community on behalf of the sustainability of Territory, rather than its colonizer and controller on behalf of the growth of Centre.

CONCLUSION: I HAVE SEEN THE ENEMY . . .

To a government that might actually seek to make such changes, the components of an ecological "industrial policy" are not difficult to envision. The elements are all there—from new job creation/business development/financing strategies, to community management boards, to more limited, oversight roles for indebted central governments.[10] Indeed, a whole "alternative" economic and political development model awaits implementation. Alas, the ultimate solution to the inertia of the liberal/administrative state lies not with the state itself, which simply cannot be expected to undertake this task without broad social support. The target for change must be our complacent social context where public groups have so far failed to provide citizens with the necessary vision of change, and leadership to get there. To date, little systematic attention has been given by forest critics as to how this might be accomplished.

This failure is admittedly a result both of the overwhelming, and pervasive, power of the Centre, and the inability of social movements to forge a new role for the state in re-forming that Centre. Instead, integral to the rise of the administrative state has been the rise of so-called "interest group pluralism".

Everything is processed through the Centre of power, so every group—environmentalist, First Nations, labour—looks to that Centre to resolve grievances. Interest groups compete for attention and for favours, with government as the referee. Thus environmentalists talk to each other, but rarely to First Nations leaders, and even less to progressive labour leaders. Without this collaborative effort, a shared vision of community dissolves, leaving an empty civic space that is quickly occupied by private economic interests which stand in overall opposition to public governance. For central governments to be "reinvented", social movements must first take back that collective context for social change.

Ecological alternatives offer the greatest potential for this reinvention because they do offer a coherent and workable path—economically, socially, and environmentally. But this ecological alternative demands that individual sectors first redefine themselves. Take, for example, organized labour. Rather than defining itself as a cohort in the defence of MacMillan Bloedel, the IWA could "re-invent" itself as a primary social vehicle for the devolution of MB tenures to community forests (staffed by former union loggers who now have residential "shares" in the local land base), and for the transformation of the manufacturing base into small businesses and worker-owned cooperatives. So too, First Nations negotiators working on treaty negotiations through an exclusivist "government-to-government" approach, could do better in the long term by fostering a larger "movement-to-movement" change. A Native island of self-government in a hostile sea of the non-Native status quo is not likely long to survive as a distinctive alternative.

This sort of analysis can be applied to the entire panoply of social interests in British Columbia forests. Environmentalists have successfully organized themselves into an effective elite group lobby (the coalition known as "BC Wild") to secure protected wilderness areas, but have not tried to move beyond that to build a broader, more systemic, movement encompassing other interests such as First Nations, labour, small business, local governments, and so on. Lacking the commitment to a broader structural change, neither has the NDP in power attempted such coalition building; quite the contrary, it has pilloried some environmentalists as "traitors", and joined with industry in mounting a multi-million dollar public relations campaign overseas to discredit its critics.

The sustainability of our forests, our communities, our planet, is in jeopardy. But, given the momentum of Centre growth, and the weight of its power, the disjointed response described here may be the best we can hope for, pending some larger catastrophes. We stand transfixed by the headlights, debating the speed (and adjusting somewhat the trajectory) of the approaching collision. Meanwhile, a turn-off and new pathway beckons, an alternative strategy makes sense *economically*, as well as environmentally and socially. How we might come to take that revolutionary turn with excitement, and without fear, is the challenge of re-invention for us all.

ENDNOTES

1   Professor of Law and Environmental Studies, and Eco-Research Chair of Environmental Law and Policy, Faculty of Law and Environmental Studies Program, University of Victoria. The author wishes to thank Cheri Burda for her research assistance in the latter stages of this paper. The financial support of the Social Sciences and Humanities Research Council is also gratefully acknowledged.

2   See "Pulpwood Fiction: the chilling, true tale of how government gave our wood away— twice", Vancouver: BC Wild, December 1994. This booklet argues that the "government is committed to providing these companies and others nearly 8.4 million cubic meters of Crown timber a year, most of it in support of high-tech, low-labour pulp mills.... And all of this wood, it should be added, is over and above today's unsustainably high rate of logging". (p.1)

3   The Commercial Back Country Recreation Policy took almost four years to prepare, coming into effect on an interim basis on June 1995. Eco-tourism schemes have been held up for years, thus delaying the implementation of local economic diversification strategies.

4   Some more high profile conflicts include Victoria, Vancouver, the Slocan Valley, and Nanaimo.

5   For a comprehensive critique of the next to final draft of the Code, see *The Forest Practices Code of British Columbia Act: A Critical Analysis of its* Provisions, Vancouver: Sierra Legal Defense Fund, 1994. As the report notes, in the Act alone, the word "despite" appears 53 times, the word "unless" 32 times, the word "exempt" 50 times. In the Standards under the Act, some of these terms appear even more frequently—"unless" appears another 77 times.

6   Not only was the Evans report shelved, but another private study was commissioned which recommended a credit system (which Evans opposed) for tenured lumber producers that supply remanufacturers or remanufacture the wood themselves.   At the time of writing, it is likely that such a credit system of benefit to the large companies will be instituted. See *Remanufacturing Supply Initiative Discussion Paper*, by Dan Perrin, Thorau Associates, February 14, 1995.

7   This attack comes from a variety of overlapping sources such as feminism, communitarianism, and Green theory.

8   See, especially the work of Robert Repetto, economist for the World Resources Institute (such as *World Enough and Time*).

9   See the Evans Report cited above.   Although small businesses offer a lower average wage than the IWA (about $35,000 vs. $53,000), in the aggregate, they generate about three times as much total wages per unit of wood as does the traditional unionized sector.

10   For example, the recent report of the Federal Standing Committee on Industry, *Taking Care of Small Business* (October 1994) is filled with innovative financing and business development strategies that are economically workable in practice and socially equitable in impact—if governments are willing to embrace this shift in scale.

REFERENCES

Best, M. (1991). *The new competition: Institutions of industrial restructuring.* Cambridge: Harvard University Press.

Frank, A.G., and Gills, A.H. (1993). World system economic cycles and hegemonial shifts in Europe: 100 BC to 1500 AD. *The Journal of European Economic History,* 22(1), 155-183.

Innis, H.A. (1956). The lumber trade in Canada. In M.Q. Innis (eds.), *Essays in Canadian economic history* (pp. 242-51). Toronto: University of Toronto Press.

Innis, H.A. (1950). *Empire and communications.* Toronto: University of Toronto Press.

Knetsch, J.L. (1995). Assumptions, behavioral findings and policy analysis. *Journal of Policy Analysis and Management,* 14(1), 68-78.

M'Gonigle, R.M., and Parfitt, B. (1994). *Forestopia: A practical guide to the new forest economy.* Madeira Park, B.C.: Harbour Publishing.

M'Gonigle, R.M. (1992). Our home and native? Creating an eco-constitution in Canada. In J. Plant and C. Plant (eds.), *Putting power in its place: Creating community control* (pp. 49-68). Gabriola Island: New Society Publishers.

O'Connor, M. (1994). *Is capitalism sustainable?: Political economy and the politics of ecology.* New York: Guildford Press.

Palmer, V. (1995). NDP cuts red tape to make sure enough wood is cut, *Vancouver Sun,* February 6, A8.

Wackernagel, M., and Rees, W. (1995). *Our ecological footprint.* Gabriola Island: New Society Publishers.

Williams, G. (1992). Greening the new Canadian political economy. *Studies in Political Economy,* Spring, 31, 5-30.

Plate 5 Winch system ➤

Plate 6 (overleaf) Hauling logs in the woods ➤

Changing Forest Practices in Coastal British Columbia

W.N. Cafferata

Chief Forester, MacMillan Bloedel

Until recently, the wealth generated for British Columbia by the coastal forest industry was reason enough for timber extraction to be socially acceptable. That has changed in recent years, and the public demand for protection of non-timber values, such as biodiversity, is clearly in the ascendancy. Industry was slow to recognize the global nature of these concerns, and tended to defend itself with arguments about economics, legal rights, and local concerns. Its critics tended to use arguments that were both global and biocentric, and posited the view of imminent and irreversible environmental destruction. Industry response to these issues changed when its customers began to express the same concerns as its critics.

Government also responded by establishing several processes to deal with the issue of land use in British Columbia, and by developing a Forest Practices Code. While the need for a code is supported in principle by environmentalists, communities, and industry, each interest has also been critical of aspects of the specific code that has been introduced. Industry's criticisms concern the effect of the Code on the cost of wood delivered to a mill, and on the amount of forest made unavailable for harvesting because of the Code. It is our view that near current levels of annual allowable cut (AAC) can be maintained while meeting full compliance with the Code.

It is also our view that for our industry's continued success in the global market, it needs to have its forest practices certified as sustainable, and it must be cost competitive with harvests that are close to today's levels.

In our opinion, government's initiative to zone forest lands according to dominant use will allow the public to better understand the costs and benefits of land use decisions, and will enable industry to practice sustainable forestry while operating to the standards of the Code. We request that government provide another key element. That is, the determination of a target AAC that is commensurate with the scale and potential of the public's forests.

BACKGROUND

The forest industry has been the major creator of wealth in coastal British Columbia since the earliest days of European settlement. Beginning with Sewell Moody's sawmill on the north shore of Burrard Inlet in 1867, and continuing through to MacMillan Bloedel's current NEXGEN paper mill development at Port Alberni, much of coastal British Columbia's social and industrial infrastructure has been created because access to timber allowed the prospect of establishing a profitable enterprise.

The right to cut the publicly owned timber came to be vested mainly in large corporations which were able to supply the capital necessary to develop the resource. Even so, most British Columbians felt that the resulting wealth was equitably distributed, because government collected substantial fees and taxes, while strong labour unions ensured that high wage scales would underpin forest-based communities.

For most of the time since European contact with British Columbia, resource exploitation was considered socially acceptable. It did not, however, command the world's attention.

That status began to change in the 1960s. In the decade after Rachel Carson published *Silent Spring*, North Americans began to notice the unforeseen and unhappy impacts of industrial civilization on fish and wildlife, on water and air quality, and on the remaining accessible wilderness. Public advocacy groups sprang up, including Vancouver's *Don't Make a Wave Committee*, formed to protest against nuclear bomb tests in the Aleutians, which later changed its name to Greenpeace. During the past 20 years, these groups have come in from the fringe to the mainstream; by 1993, according to *Outside* magazine, the 10 largest non-government environmental organizations in America had a combined membership of over six million, and their budgets totalled over $500 million U.S.

FOREST PRESERVATION—A GLOBAL DEBATE

Preserving old-growth forests from commercial exploitation was a key focus of these organizations by the 1980s, when controversies over forest land use began to flare up in coastal British Columbia. One of the earliest concerned the Tsitika Valley, where logging had been planned to proceed in 1973. The plans met with strenuous objections by people who wanted the entire watershed preserved, leading to intensive discussions at the local and provincial levels, which in 1978 resulted in a compromise land-use plan. That agreement lasted, with some occasional upsets, until it was subsumed by the Commission on Resources and the Environment (CORE) process in 1993.

A number of other similar controversies followed, marked by a pronounced shift in the strategies of those who were campaigning to preserve coastal forests from industrial use. The strategic shift was foreshadowed in the struggle to establish a park on South Moresby Island in the Queen Charlottes. That dispute was initially a local matter, with strong support in the local community for continued logging. But the campaign was soon extended to national and international forums. Local interest groups and First Nations mobilized political support in Ottawa, in Europe, in Washington, D.C., and at the United Nations. They scored a victory in capturing the world's attention when they were able to bring former U.S. President Jimmy Carter to South Moresby, where he declared to the media that he thought the area should be a park. Shortly thereafter, Ottawa designated 147,000 hectares of South Moresby, approximately 15 percent of Queen Charlotte Islands, as a park reserve.

The British Columbia forest industry was slow to realize the implications of the shift from the local to the international arena. The industry continued to rely on arguments that had stood the test of time and which resounded most effectively among British Columbians: the value to the province of industry-generated jobs and tax revenues; the industry's legal right to develop the resources under its tenure; the priority of humankind's need for forest products over the needs of other species; the efforts at reforestation and protecting fish habitat.

The industry's arguments, reflecting the views of its leadership, tended to be locally based and anthropocentric; the opposition's thrust was global and biocentric. They demanded that the forests be preserved so that the world could fend off irreversible environmental destruction: global warming, depletion of the ozone layer, extinction of species, loss of biodiversity. Within that context, the industry's arguments paled to insignificance.

THE INDUSTRY RESPONDS TO CHANGE

In recent years, the industry has broadened its defence to include recognition of non-timber values, while changing its forest practices accordingly. Public opinion poll results indicate that these actions have improved the industry's credibility within British Columbia and Canada. However British Columbia's forests sector has little support outside the province, as is evident in the continuing campaigns by Greenpeace Germany, Greenpeace UK, the Rainforest Action Group, and other groups that seek to pressure the British Columbia government to reverse its decision to allow logging in parts of Clayoquot Sound.

The most recent recipients of messages from these overseas groups have been an audience of great importance to MacMillan Bloedel—our customers.

The message has been delivered in various ways. In the United Kingdom, our customers received threats that their products, made from material provided by MacMillan Bloedel, would be publicly attacked as "environmentally incorrect." In Germany, the threat was one of boycotts. In California, protesters chained themselves to the gates of factories that used our products. In each case, the customers being threatened were also presented with well prepared texts and photographs purported to support allegations of large-scale and irreversible destruction of British Columbia's forests.

The customers' response was to demand information from MB. Were the allegations about massive non-compliance with environmental regulations true? Were British Columbia forest companies completely unregulated? What was the state of forest practices in British Columbia? And, most of all, what was the industry doing to gain public confidence in its practices?

Not surprisingly, the industry is now making a much greater effort to supply international customers with concise and accurate information to counter the accusations made by those who seek to halt commercial use of British Columbia's coastal temperate forests.

GOVERNMENT RESPONSES

In the 1990s, the temperate rain forest is the focus of international environmental campaigns. But before advocacy groups targeted coastal British Columbia, tropical forest management had come under similar scrutiny. In response, the major producers of timber in Southeast Asia formed the International Tropical Timber Organization (ITTO) in 1983, and committed themselves to a set of principles of sustainable forestry.

The shift of focus from tropical to temperate forests was quite apparent by the time of the 1992 conference of the United Nations Commission on Environment and Development in Rio de Janeiro. The conference produced an international convention for the protection of biodiversity, which Canada signed. Since then biodiversity has become, in the eyes of critics of managed forests, the essential component of sustainable forestry.

While the UNCED convention is general in nature, Washington State has established a detailed set of rules for sustainable forestry built around maintaining biodiversity. These rules, which have some relevance to conditions in coastal British Columbia, came into effect just as the British Columbia coast forests began to attract international notice. Among the salient aspects of the Washington code are provisions for: wildlife reserve trees; wetlands; size and timing of clearcuts; stream temperatures; critical habitats; use of chemicals; special forest practices; and cumulative effects.

British Columbia's response to the increasing international sensitivity is the new Forest Practices Code, which is to be enforced by strict regulation, and which is expected to be proclaimed into law during 1995. The new Code stresses non-timber values, respect for traditional aboriginal uses of the forests, public involvement in forest planning and protection of biodiversity. As previously stated, it is supported in principle by environmental groups, communities and industry, although each has criticized some of its provisions.

THE COSTS OF THE CODE

Perhaps predictably, industry's criticisms of the Code have mainly related to its effects on the cost of wood delivered to a mill, and on the amount of wood that may be harvested.

MacMillan Bloedel estimates that the Forest Practices Code will add about 7 dollars per cubic metre to the costs of logs delivered to a mill. This represents a greater than 10 percent increase over 1993 costs. Whether this is too much, too little, or of no concern at all depends entirely on one's point of view. But the industry considers it worth noting that its return on capital employed from 1987 to 1993 was less than 7 percent (Price Waterhouse).

It is also instructive to examine the total economic value available from the timber portion of the forest, apportioned among government, labour, and industry. Figure 4.1 represents the difference, over 30 years, between the selling price of logs on the British Columbia coast and the cost of logging, before accounting for stumpage, royalties, depletion and return on capital employed. The irregular lines represent yearly differences, and the smooth lines represent trends.

The trend lines indicate a constant margin of about 20 dollars per cubic metre, to be shared among the three parties. At the moment, the industry is enjoying high prices for its products (although stumpage has risen appropriately), and can absorb the extra operating costs imposed by the new Code. However, when the market inevitably turns down, those costs will be felt. Indeed, if 7 dollars per cubic metre had been added to costs between 1982 and 1992, it would have wiped out any industry profit, even if stumpage had been set at zero.

Beyond the impact on operating costs, the industry is concerned about the Code's effects on volumes of wood available for harvest. Analysts outside the industry have estimated that the Code will reduce harvest levels by 10 to 20 percent, and have forecast an annual cost to the British Columbia economy of between 500 million dollars[1] and 1.5 billion dollars.

The industry contends, based on a macro-view of the commercial forest land base, that there is room for sustainable forestry and substantial harvesting of timber. The numbers in Table 4.1 assume the commercial forest land base on the coast would produce timber at rates equal to 80 percent of fully stocked natural stand yields. The coastal land base is estimated to be approximately

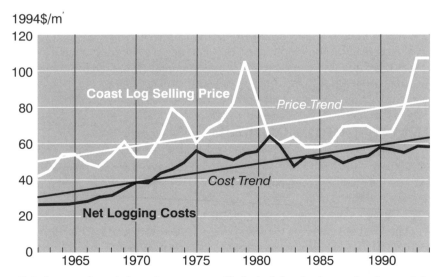

Net of economic rents (e.g. stumpage, royalties), depletion & return on logging capital.
Sources: COFI, MB, Price Waterhouse

Figure 4.1   A comparison of the selling price of logs, on the B.C. coast, and logging costs

15 million hectares, of which 9.7 million hectares are described as forested, and roughly 5.0 million of those hectares are classed as commercially accessible. If common ground can be found among the forest stakeholders, a sustainable coastal cut of 24 million cubic metres per year is achievable. Of course, if we cut the wood, we'll also want to sell the wood. This brings us to certification.

Table 4.1   Present AAC levels can be maintained

	Commercial Land Base million ha	MAI m³/ha/yr	Possible AAC million m³/yr
Coast	3	8	24
Interior	17	3	51
Province	20		75

CERTIFICATION

The international effort to preserve temperate coastal forests has evoked another level of response by industries and regional and national governments: the campaign to create international standards for certifying forest products that are derived from sustainably managed forests. In addition to the above-mentioned ITTO and Washington State rules for sustainable forestry, a number of other certification systems are in place or in process. Some of them are put forward by international environmental non-government organizations. Canada is well advanced in developing national standards for sustainable forestry through the Canadian Pulp and Paper Association and the Canadian Standards Association (CSA); the standards adopted by Canada will be submitted to the International Standards Organization for their approval.

The forest industry recognizes that the credibility of any standards will depend on their acceptance by reputable environmental organizations. An onerous criterion for certification of sustainable forest practices by such organizations will be the guarantee of a "chain of custody"—a manufacturer must be able to demonstrate that all wood supplied to its mill comes from sustainably managed forests. In British Columbia, backtracking each piece of wood from a converting plant to its place of origin would be nearly impossible; the industry believes that this requirement can be met through compliance with the B.C. Forest Practices Code, and the CSA standards, as verified by independent audits. Since the Ministry of Forests' timber sale program makes it the largest harvester of wood in the province, its harvest operations may have to be audited by a third party if the chain of custody requirement is to be met.

WORLD WOOD DEMAND/SUPPLY

The British Columbia forest sector economy is built upon the world's demand for its products. A key question in the future of that economy is: what is the world willing to pay for what the British Columbia forest industry produces?

Some suggest that the world demand for wood is growing faster than the available supply. This leads to the view that real prices for wood will increase, and that we in British Columbia can continue to enjoy our enviable standard of living even though we reduce our harvest well below current levels. So let us examine the underlying assumptions required for this vision of the future to be realized.

First, let me restrict my remarks to industrial roundwood as opposed to fuel wood. The two types of wood use are not usually competitive with each other. Also, wood is much more valuable if it can be used for industrial purposes than as fuel.

If we are to boost demand significantly we must assume that future world population will consume industrial wood at a similar rate as the present and recent past inhabitants of planet earth. This is unlikely to be true. Annual per capita wood consumption in North America, Europe and Japan, regions with almost no population growth, is 10 times higher (at 1 m^3/person) than it is in the rapidly growing population areas of the world (at 0.1 m^3/person per year).

So population growth is not going to push global wood consumption up at rates comparable to average world per capita consumption. Even as those countries that have had low wood consumption rates and low incomes do get richer, they may not use much more wood as it may not be part of their cultural pattern to do so.

Then we must look at the supply expansion potential for the world's plantations. In 1994, just two countries (Chile and New Zealand) planted 250,000 new hectares of forest[2] that will grow about five million cubic metres of industrial roundwood a year. This is enough wood to satisfy the needs of 50 million people in countries without a wood-using culture. Current new planting is much more than this on a global basis and although it will be 10, 20, or 30 years before the yields are available, sufficient expertise and profitability has been attained by local tree growers that the trend seems bound to accelerate. Austral-Asia and the southern part of South America seem particularly suited to growing trees. Any temporary real increase in timber prices will just encourage more plantations.

Finally, if we are to have real price increases we must assume limits on substitution, conservation, and wood extending techniques. We don't have to look far to see innumerable examples of these. A few examples:

- steel studs for wooden; parallel strand lumber for solid beams;

- improved recovery in sawmills; thin veneer overlays on cheap substrate; thinner and lighter newsprint;

- greatly improved reuse and recycle of non-wood material. (For a good article on this see "Manufacturing for Reuse"; Gene Bylinsky; *Fortune Magazine*, February 6, 1995.)

We have concluded that the world is not short of industrial wood and that there is unlikely to be more than temporary, regional spikes in price while the counteracting economic forces do their moderating work.

An exception to these conclusions is seen in the world of appearance grades of lumber and logs where the commodity price is set per square foot of appearance effect. With ever thinner veneers giving the appearance effect and better technology extracting more appearance grade material from old growth logs, a real price increase in coastal old growth timber has been observed. Lately this has been aided by the devaluation of the Canadian dollar. We expect all three of these trends to moderate as limits to technology are approached.

So we, and most of the analytical observers of the world forestry scene, believe that real wood prices will not rise over the long haul.

That is why we are persistent in our argument that British Columbia's harvest level be maintained at close to today's levels, and that our forest industry be cost competitive. And we believe this can be done in a manner consistent with the principles of sustainable development.

But what if we are wrong about there not being a real price increase for wood? Then British Columbia would have a handsome windfall profit, and its government would have to deal with disbursing the resulting high economic rents.

But if British Columbia plans as though it were inevitable that a real price rise would compensate for high log costs and a reduced harvest level, and it doesn't happen, then not only has the present generation of workers been deprived by reductions in forest output, but so have future generations.

Before we make a decision to risk our standard of living on the premise of rising real prices for wood, we would be wise to do some hard analysis of the costs and benefits of such a course.

CONCLUSIONS

If British Columbia's coastal forest industry is facing a combination of increased costs and slow growth in world demand for wood products, the forest sector's best hope—and that of the economy it sustains—is for the industry to become both highly efficient and fully sustainable. Happily, both goals are achievable. If it transpires that world demand for wood does not level off, the benefits to the province—and particularly to its rural inhabitants—will be so much the greater.

British Columbia can meet any reasonable international standard for certification as a practitioner of sustainable forest management, while preserving 12 percent of its land base in legally protected areas outside the commercial forest, and keeping at least that much additional forested land in minimal contact with human activities.

The problem to be addressed in the coastal forests of British Columbia is not a matter of irreversible damage to ecosystems or biodiversity. It is arguably not an environmental problem at all, but a social and political problem. Its solution will begin with the recognition that not every hectare of forest land can satisfy every demand that every interest might wish to place upon it. To pretend otherwise, as some do, is to invite inevitable social disruption and environmental disruption.

The tools to deal with this problem are at hand. For example, I think that the work being done to put in place the Vancouver Island Land Use Plan is an excellent opportunity to identify common ground and build upon it. Zoning of forest lands in accordance with their dominant use will allow society at large to better understand where the compromises between conflicting values are made, and what the costs and benefits are of these land use decisions. Adherence to the standards of the Forest Practice Code in all zones will provide for sustainable forest practices. Similar work is going on in the Cariboo and other regions of British Columbia.

We ask that government listen to the requests of both the forest industry and forest dependent communities of British Columbia to provide another key element for the transition to sustainable forestry in British Columbia. That is the determination of a target annual allowable cut that is commensurate with the scale and potential of the public's forests, and regulations that enable its efficient achievement.

ENDNOTES

[1]   van Kooten, G.C. (1994). Cost-benefit analysis of B.C.'s proposed Forest Practices Code, April 27, 1994.

[2]   Sedjo, R.A. (1995). The potential of high-yield plantation forestry for meeting timber needs: Recent performance and future potentials; Discussion Paper 95-08.

Plate 7   Automation in the pulp and paper industry   ➤
Plate 8 (overleaf)   A paper machine: wet end   ➤

Forestry Land Use and Public Policy in British Columbia: The Dynamics of Change

5

Thomas Gunton

School of Resource and Environmental Management,
Simon Fraser University

In the Spring of 1994 over 30,000 people gathered on the steps of the legislative building to protest against proposed land use changes on Vancouver Island. This demonstration, one of the largest in British Columbia history, reflects the intensity which has characterized provincial land use policy for decades. Yet despite this conflict between resource extraction and protection, the provincial government recently implemented significant changes in land use which have been largely accepted by most British Columbians, including the forest industry.

The purpose of this chapter is to explain how these changes were made with such a high degree of support after decades of intense conflict. The chapter begins with an outline of the staple theory as a framework for understanding resource policy. Recent changes in provincial resource policy are then summarized and explained in the context of staple theory.

STAPLE THEORY

The development of Canada has been understood in terms of the need for extraction of a succession of staple products for export to meet the requirements of more advanced industrial nations (Aitken, 1959; Bertram, 1967; Watkins, 1963). Following Bertram, staple products are defined as being based on natural resource extraction involving little processing prior to export (Bertram, 1967: 75). External demand for the staple product generates economic expansion by the direct investment necessary to extract the staple and by the spread effects stimulated by investment in further processing of the staple prior to export (forward linkages), manufacture of inputs required in the staple sector (backward linkages), and provision of services to those employed in the staple

sector (final demand linkages). Growth can be further stimulated by generation of resource rents defined as a return to the staple above normal returns to
labour and capital.

Given its powerful role in economic expansion, the staple sector dominates both private and public decision making. To quote Innis, one of Canada's
leading economic historians:

> *Energy has been directed toward the exploitation of staple products
> and the tendency has been cumulative. Energy in the colony was
> drawn into the production of the staple commodity both directly
> and indirectly. Population was involved directly in the production
> of staple and indirectly in the production of facilities promoting
> production. Agriculture, industry, transportation, trade, finance,
> and governmental activities tend to become subordinate to the pro-
> duction of the staple for a more highly specialized manufacturing
> community* (Innis, 1930).

And each staple has its own requirements and impacts and the shift to
new staples can create periods of disequilibrium as institutions adjust to new
demands. Again from Innis:

> *Concentration of the production of staples for export to more highly
> industrialized area in Europe and later in the United States had
> broad implications for the Canadian economic, political and social
> structure. Each staple in its turn left its stamp and the shift to new
> staples invariably produced periods of crises in which adjustments
> in the old structure were painfully made and a new pattern created
> in relation to a new staple* (Innis, 1950).

Not surprisingly, the evolution of resource dependent British Columbia
has been extensively analysed within the context of staple theory (Gunton, 1982;
Marchak, 1983). Economic expansion has been largely driven by the forest,
mining and fisheries sector for much of British Columbia's history and public
policy has been directed towards facilitating this expansion by favourable property rights, and public investment. In this environment, land use policy has
been designed to fulfil the requirements of the forest sector. Over 83 percent of
the land base has been designated as provincial forest to meet the needs of
forest harvesting (B.C. Ministry of Environment, 1996: 4). Tenure rights on this
land have been designed to provide long term security to facilitate investment.
Non-extractive uses such as wildlife have rarely been provided legal protection by the province so as to not interfere with timber harvesting. Expansion
of protected land such as parks that impacts on the forest and mining sector
has been strongly resisted and attempts to negotiate treaties with First nations
that transfer property rights from the resource sector have not been successful.
This direction in public policy is of course consistent with staple theory.

RECENT CHANGES

In the last several years there has been a profound change in the nature of land and resource policy. Protected areas have been significantly expanded at the expense of forestry and mining, legal obligations to protect non-extractive uses have been strengthened, resource pricing policy has changed, and treaty negotiations with First Nations have accelerated.

The expansion of protected areas is summarized in Table 5.1. The data show that the pace of expansion increased in the period 1990-1995 with the addition of 2.5 million hectares to the provincial protected area system. The commitment is to have, by the turn of the century, 12 percent of the land base in protected area status where timber, mining, and other resource extractive activities are prohibited. In the Spring of 1995, 108 new parks were created and given legislative protection. The effect has been to reduce timber harvest by about 3 percent during a period when the Annual Allowable Cut (AAC) is already declining as a result of lower estimates of the long run sustained yield.

Table 5.1   Protected area in British Columbia

Year	Hectares	Provincial Land Base (%)	Increase From Previous Years (%)
1975	4 569 401	5.0	
1980	4 592 256	5.0	1
1985	4 750 432	5.0	3
1990	6 139 070	6.5	29
1995	8 669 500	9.2	41

Source: B.C. Ministry of Environment, Lands and Parks, Unpublished data base.

A second major change is the enactment of the new Forest Practices Code which strengthens the legal commitment to protect non-timber values. The new Code gives the force of law to previous guidelines to protect fish and wildlife. It provides, for the first time, legal requirements for extensive set-backs along rivers, reduced size of clearcuts, requirements for set-asides to protect wildlife and scenic views and new low intensity road building and harvesting techniques to protect soils and slopes. The Code is enforced by new institutional structures such as the Forest Practices Board, which monitor compliance and hear public complaints and a substantial increase in fines.

Like the expansion of protected areas, the new Forest Practices Code will have a profound effect on the forest sector. The Code is expected to raise harvesting costs for the forest industry by approximately $300 to $500 million per year or up to 10 percent (Saunders, 1993). The increased protection for non-timber values is expected to reduce the AAC by a further 6 percent (Price Water-house, 1995). The combination of protected area expansion, the Forest Practices Code, and timber supply reviews designed to reduce AAC to long run sustained yield could result in a total decline in AAC of up to 17 percent (Price Waterhouse, 1995/96: 66). Although this reduction may be compensated by other measures, such as more intensive silviculture, they will result in the first significant decline in the AAC in the postwar period (see Figure 5.1).[1]

Harvest Levels (m³)

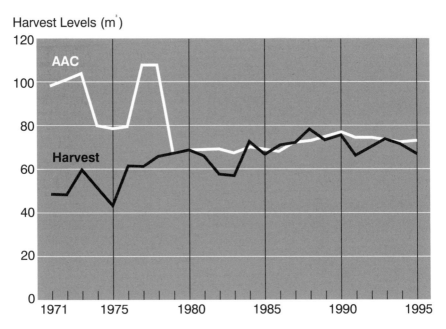

Figure 5.1   Annual harvest levels compared with AAC's on Crown Lands, 1971-1995

A third change was the creation of Forest Renewal, a crown corporation designed to invest in the forest sector. Forest renewal is funded by a significant increase in the price that the crown charges private companies for timber. The charge is based on an *ad valorem* charge on timber which is anticipated to raise an additional $400 to $500 million per year, representing an increase of over 50 percent in stumpage paid. Like the changes in land use, this increase has a profound impact on the forest sector by reducing profits of the forest companies. According to one recent analysis, the increase in timber charges has cost the forest companies up to $3 billion in terms of reduced asset value (Binkley and Zhang, 1995). Yet, surprisingly, this change in timber charges was supported by much of the forest sector.

A final change is the acceleration of treaty talks. The provincial government has concluded a cost sharing agreement with the federal government and has committed to a comprehensive treaty process. After decades of opposition, the province is now an active proponent seeking resolution of treaties with an estimated public cost of approximately $6 billion (KPMG, 1996).

DYNAMICS OF CHANGE

The changes described above cumulatively have resulted in fundamental change in the management of the largest staple industry in British Columbia. In apparent conflict with the precepts of staple theory, the state has intervened in a manner to reduce the rate of harvest, increase the cost and lower profits of the forest sector. And the industry has either supported or accepted these changes with little opposition. These changes clearly raise doubts about the predictive power of staple theory in explaining public policy.

On closer examination, however, these changes in policy are fully consistent with the staple theory framework. As Innis suggests, each staple leaves its own stamp, and shifts to new staples precipitate transitions to accommodate the needs of the new staple sector. The staple sector in the British Columbia economy has clearly been going through a transition. The forest industry's role in the British Columbia economy has declined significantly from 13 percent in 1970 to 8 percent in 1994 (B.C. Economic Review, 1995). The tourism sector represents about 5 percent of the economy, and the component dependent on natural resources or ecotourism has grown in relative importance. The needs of ecotourist industries such as recreational fishing, skiing and hiking are contrary to the requirements of the forest sector. Land must be left in its natural state, ecological diversity maintained, and fish and wildlife resources protected. The industries in this sector, therefore, provide a powerful lobby in support of changes in forest practices, expansion of protected areas and regulations protecting wildlife.

While the declining role of the forest sector and the growing strength of the emerging staple-based ecotourist sector is an important factor explaining these changes, it is not sufficient. The new staple sector is still relatively small compared to the forest sector and some of the changes, such as the increase in timber charges or treaty negotiations, are not policy adjustments promoted by the ecotourist sector. Therefore other factors must be involved. Following Innis, the other major factor must be that the changes are somehow in the interests of the dominant staple industry, forestry.

This is indeed the case. The forest sector in British Columbia is an export-based industry dependent on access to foreign markets. Two major impediments to foreign markets arose in the last several years which have threatened the viability of the forest sector. First, the American government at the behest of the United States forest industry has launched a series of countervail actions to restrict imports of British Columbia timber into the United States. A countervail was initiated in 1986 and resulted in the imposition of an import levy of approximately 15 percent on all imports (Percy and Yoder, 1987). The levy was justified on the grounds that the British Columbia forest sector had an unfair advantage over the United States sector because the government subsidized its forest industry by charging less than the fair market price for public timber sold to private companies. The levy caused substantial harm to the British Columbia industry and, through an agreement negotiated between the American and Canadian governments, was replaced in December, 1986 with an increase in the stumpage payments to the provincial governments equivalent to the amount of the American levy. Although this did not reduce the cost to the timber industry, it did ensure that the increased payments remained in British Columbia instead of being collected by the United States government.

The agreement was terminated unilaterally by Canada in 1991, and the American government retaliated by again imposing an import levy on the grounds that the Canadian industry was subsidized. Discussions on Forest Renewal and the increase in stumpage occurred between the British Columbia government and the industry during the imposition of this second countervail levy. One factor affecting the discussions was the view that an increase in the stumpage rate by the provincial government would increase the probability of the United States government cancelling its countervail. The United States complied in 1994 when it dropped its levy.

Another factor reducing resistance to the increase in the provincial stumpage rate was that the industry was enjoying high profits and high prices during the negotiations on forest renewal and the provincial government was considering an increase in stumpage charges to address serious fiscal problems. From the industry perspective, a second best option was to have the increase in stumpage placed in a special fund to be reinvested back into the forest sector instead of having the increase flow into general revenue to be used for other purposes. The combination of warding off United States Countervail and preventing the loss of stumpage revenue to uses other than forestry convinced the more astute members of the forest sector to support an increase in stumpage to fund Forest Renewal.

A second impediment to market access was the growing threat of boycotts by environmentalists to protest forestry practices in British Columbia. In Europe, opposition to imports of British Columbia forest products had grown to the point where the European governments were considering banning such imports. In Germany, polls showed that the majority of the public supported a ban. In the United States, similar opposition was developing to pressure purchasers of British Columbia forest products to switch to alternative suppliers.

These threats helped convince the forest industry of the need for a counter campaign, the credibility of which depended on implementing enough changes in British Columbia forest practices to convince foreign markets that a boycott action was unwarranted. The implementation of the new Forest Practices Code and the expansion of protected areas were accepted by much of the industry as necessary changes to avert boycotts and improve the international reputation of the British Columbia industry. The industry, in fact, utilized some of these changes in an international advertising program. In this environment, opposition to the changes was unwise and uncommon.

Another factor promoting change was the growing public opposition to forest practices within the province itself. Fuelled by pressure from the ecotourist sector, the British Columbia public was demanding changes in forest practices and land use. Public appetite for major changes against an increasingly unpopular forest sector no doubt convinced the industry that some changes were necessary to avert even more significant changes, such as an overhaul of the tenure system which had been advocated in many reports (Pearse, 1975; Forest Resources Commission, 1991), and would seriously undermine the interests of the larger companies.

Treaty discussions with First Nations were also initiated, in part by the needs of the resource sector for certainty in property rights. Several recent court cases in British Columbia have influenced the existence of First Nations' aboriginal rights in land use management (Morgan and Thompson, 1992). This has led to growing legal challenges and civil disobedience which have disrupted harvesting activities. The reduction of uncertainty of property rights by resolution of land claims is important to investors in the resource sector.

The changes in public policy proved to be successful for the forest sector. United States countervail was eliminated in 1994, and the provincial government did not transfer any of the increased stumpage charges into general revenue. The threat of foreign boycotts has declined and the credibility of the industry has improved significantly. The credibility of the industry with the British Columbia public now slightly exceeds the credibility of the environmental movement, a remarkable change in just a few years (Markettrend, 1995). This has helped avert even more threatening changes, such as major tenure reform.

Although most of the changes ultimately benefited the forest industry, they were not accepted without some resistance. Therefore, a final factor explaining these changes was the strong commitment by the provincial government to reform the forest sector. This motivation in part reflected the public support for change and the public position articulated by the government prior to its election. This commitment to change on the part of government was a necessary condition for overcoming this resistance within the forest sector and exploiting the opportunity for reform.

CONCLUSION

Staple theory suggests that public policy is heavily influenced by the needs of the dominant staple industries which drive economic growth. While recent changes in provincial policy in British Columbia may appear contrary to this proposition, they are in fact consistent with the theory. The increasing role of the new ecotourist sector combined with the decline of the relative signifi- cance of the forest sector has required some important changes to protect natu- ral resources. These changes have also been necessary to protect the long run viability of the forest industry by ensuring access to foreign markets and re- investment of increased stumpage charges back into the industry. Although these changes were not accepted by industry without some tension and adept management by the provincial government, they were implemented with sur- prisingly little controversy. This coinciding of enlightened self interest on the part of resource industries and enlightened public policy making on the part of government has resulted in fundamental changes in resource policy which will improve long run sustainability and social welfare.

ENDNOTE

[1]   Although the AAC has declined in other years, such as 1979, these declines did not result in reduced cut because the AAC was not fully utilized.

REFERENCES

Aitken, H.G.J. (1959). Defensive expansionism: The state and economic growth in Canada. In W.T. Easterbrook and M.H. Watkins (eds.), *Approaches to Canadian economic history*, (pp. 183-221). Toronto: McClelland and Stewart.

Bertram, G.W. (1967). Economic growth in Canadian industry 1987-1915: The staple model. In W.T. Easterbrook and M.H. Watkins (eds.), *Approaches to Canadian economic history*, (pp. 74-98). Toronto: McClelland and Stewart.

Binkley, C.S., and Zhang, D. (1995). The impact of timber-fee increases on B.C. forest products companies. April, 1995. Unpublished.

British Columbia Provincial Government (1994). Bill 40—Forest Practices Code of British Co- lumbia Act.

British Columbia Provincial Government. British Columbia Land Statistics, May, 1985. Min- istry of Lands, Parks and Housing.

British Columbia Provincial Government. The Second Annual Report of the British Columbia Treaty Commission for the Year 1994-95.

Forest Resources Commission (1991). Background Papers. Victoria: Forest Resources Com- mission.

Gunton, T. (1982). *Resources, regional development and public policy: A case study of British Colum- bia*. Paper #8. Ottawa: Centre for Policy Alternatives.

H & W Saunders Associates Ltd. (1993). A proposed Forest Practices Code for British Columbia: Background Papers.

Innis, H.A. (1956). *The fur trade in Canada: An introduction to Canadian economic history* (2nd ed.). New Haven: Yale University Press.

Innis, H.A. (1950). *Empire and communications.* Toronto: Ryerson.

KPMG (1996). Project report: Benefits and costs of treaty settlements in British Columbia—A financial and economic perspective. January, 1996.

Marchak, P. (1983). *Green gold: The forest industry in British Columbia.* Vancouver, B.C.: University of British Columbia Press.

Morgan, N.A., and Thompson, A.R. (1992). Crown lands and native land claims: British Columbia challenges and opportunities. *Environments: A Journal of Interdisciplinary Studies,* 21, 17-25.

Pearse, P. (1975). *Timber rights and forest policy in British Columbia.* 2 vols. Royal Commission on Forest Resources. Victoria: Queens Printer.

Percy, M.B., and Yoder, C. (1987). *The softwood lumber dispute and Canada—US trade in natural resources.* Halifax: Institute for Research on Public Policy.

Price Waterhouse (1995). Forest Alliance of British Columbia: Analysis of recent British Columbia Government forest policy and land use initiatives. September, 1995.

Watkins, M.H. (1969). A staple theory of economic growth. *Canadian Journal of Economics and Political Science,* XXIX(2), 141-158.

Plate 9 (overleaf) Lumber: ready for export ➤

Implementing Forest Policy Change in British Columbia: Comparing the Experiences of the NDP Governments of 1972-75 and 1991-?

6

Jeremy Wilson

Department of Political Science, University of Victoria

The British Columbia forest sector has experienced continual flux since the 1960s. The practices, structures, rights, and obligations of industry and state components of the forest sector have undergone major changes, and powerful new actors have elbowed their way into the policy making process. One good indicator of the degree of change is the long list of new concepts that have become prominent in forest policy discourse in the last 25 or 30 years; the appearance of concepts like falldown, integrated resource management, sympathetic administration, old growth, biodiversity, endangered species, stakeholder, shared decision making, and a host of others reflects the extent to which old operating routines and ideas have been challenged and discarded.

Each of the currents of change which run through this sea of flux has complex origins. Some are primarily the result of exogenous developments like the rise of American countervail pressure. Others stem mainly from made-in-B.C. decisions by industry actors, while others are largely the product of government policy change.

This chapter focuses on change resulting from shifts in government policy. It reflects on why policy ideas like those advanced by numerous contributors to this volume do or do not translate into policy change. In an attempt to illuminate some of the factors which help or hinder government efforts to design and implement forest policy change, the experiences of British Columbia's two New Democratic Party (NDP) governments, the Harcourt Government elected in October 1991, and the busy but short-lived Barrett Government of 1972-75, are compared. Both were, broadly speaking, committed to forest policy change; but their reform agendas, their approaches to achieving change, and their rates of success were very different.

The chapter's conclusions derive from two short narratives. The first focuses on the Barrett Government's natural resources czar, Bob Williams. We reflect on Williams' inability to translate quite radical inclinations into significant change. The second story focuses on the Harcourt Government's pursuit of a reform agenda comprised of an ambitious list of sustainability and industry transition initiatives. We reflect on its choice of moderate reform priorities and on its rejection of a more radical course.

THE SOCIALIST BOGEYMAN VS. THE DINOSAURS

Bob Williams' tenure as forests minister was an active one. In an attempt to preserve jobs and forestall increased industry concentration, he brought a number of companies under public ownership. He tried to increase stumpage and royalties, and intervened strongly in the chip marketing process. He established the Environment and Land Use Committee (ELUC) Secretariat, thus improving inter-agency coordination and increasing the bureaucracy's regional planning capacities. He bolstered the Fish and Wildlife Branch's capacity to influence forest planning, established the Purcell Wilderness Conservancy, and placed logging moratoria on preservation candidate areas such as the Stein and the Valhallas.

This was an impressive list of accomplishments, but measured against the vision of a reformed forest economy presented by Williams prior to the 1972 election, this list must be considered disappointing. Williams was above all a "rent collector", concerned that British Columbia get "a better shake on our resources". According to his analysis, too much economic rent was being creamed out of the forests by large, inefficient corporations enjoying the dual advantage of low stumpage prices and government-granted regional monopoly positions (Williams, 1971). This non-competitive tenure system should be replaced by a diverse range of management and ownership forms. The desirable mix would include some publicly owned enterprises operating under regionalized or community-based forms of popular control, but public ownership was not presented as a panacea. An NDP government, said Williams, should try to promote a diversified private ownership sector which, among other elements, would include thousands of small "family scale" forest farmers. According to a background paper to the 1971 convention, "a broad 'people power' leasehold private ownership sector in the forest industry might well be the most effective means of 'socializing' the industry" (Williams, 1971a).

Any attempt to explain Williams' failure to achieve more of this agenda must start by noting that his vision of a more diverse forest economy was not backed up by a clear blueprint for achieving change. Williams brought to office a well developed critique, a tough-minded disdain for many of the orthodoxies that had legitimated industry and government practices during the W.A.C. Bennett years, and a broad vision of a different kind of forest economy. He did not bring a clear plan for achieving that vision. A comprehensive blueprint might have been developed through an early royal commission, but this idea was apparently not seriously considered. Instead, after giving himself 30 days to explore the territory, Williams launched into action on a variety of fronts, banking on the hope that through hard work he could integrate various initiatives into a coherent change agenda and guide the process forward.

Williams did not totally ignore the importance of laying the foundation for fundamental change. Most importantly, the Task Force on Crown Timber Disposal, chaired by Peter Pearse and appointed in early 1974, was intended to prepare the way for rent collection initiatives. From the outset, however, much of Williams attention was focused on reacting to unfolding developments. As he encountered more and more obstacles, he increasingly came to believe that the best way to achieve change was to give bright, "feet on the ground" people room to experiment with locally appropriate solutions. However, he also reversed an early decision not to have a royal commission, finally handing the task of developing a change blueprint to one expert he did apparently trust, University of British Columbia forest economist Peter Pearse. As it turned out, the election was called and lost before Pearse reported.

Let us survey these events in more detail.

Tenure and Stumpage: Exploring the Limits of the Possibilist Terrain

Early in the new government's term Williams warned the coastal "dinosaurs" that the "former peanut pricing for the commodities the people of this province own is not going to continue", and spoke of the need to "trim the fat" from the coastal industry (*Vancouver Sun*, December 12, 1973: 1). He clearly looked to the Pearse Task Force for advice on how the government might collect more rent from the industry. It was asked to formulate recommendations on royalties, stumpage charges, and tenures with a view to ensuring "that the full potential contribution of the public forests to the economic and social welfare of British Columbians is realized . . ." (B.C. Task Force on Crown Timber Disposal, February 1974: 51). Williams was satisfied with the advice. The task force quickly produced a first report on arrangements governing what were referred to as "old temporary tenures," the various leases and licenses granted before 1907 that covered about 2 million acres of British Columbia's best timber land, and were held for the most part by the four or five largest companies. In what Williams termed an "exciting, excellent document", the task force recommended that the present schedule of fixed royalties be replaced with a regular stumpage rate system (*Vancouver Sun*, March 7, 1974: 34). It was estimated that the changes suggested would produce a net increase in provincial revenues of over 30 million dollars. Legislation implementing these changes was passed in 1974, but the decline in world lumber markets which began in 1974 apparently convinced Williams that the industry should not have to bear more costs. The legislation was not proclaimed into law (Williams, 1981: 38-39).

In its second report, released in July 1974, the Task Force turned its attention to the system for appraising the value of Crown timber. The report stressed that, in the absence of competitive timber sales, the appraisal system fixed the price paid. It thus bore ". . . the full brunt of protecting the public financial interest in timber resources" (B.C. Task Force on Crown Timber Disposal, July 1974: 151). This report confirmed Williams' views about the consequences of the lack of competition for timber, concluding "that the present arrangements governing the marketing of logs, pulp chips and other intermediate timber products are not adequate to protect the public interest in Crown timber" (p. 127). Neither the coastal log market nor the interior pulp chip market could be depended on to yield prices that reflected the full value of Crown timber (pp. 127-28). To improve the markets' performance and establish a reasonable basis for appraisals, the task force recommended that a new Crown agency be mandated to participate as a middleman in intermediate forest products markets. The functions of the proposed Timber Authority would include: "standing ready to purchase logs available for sale in competition with other buyers; maintaining facilities for handling, sorting and storing logs where their value can be enhanced by these means; and selling logs competitively . . ." (p. 128).

In late 1974 Williams planted a version of the Timber Authority idea in legislation giving the cabinet the power to raise the prices pulp mills paid sawmills for chips. The Timber Products Stabilization Act went beyond the interior chip market problem to establish a public corporation, the B.C. Forest Products Board. This board was to have broad powers ". . . to improve the performance of markets for forest products, and to encourage the utilization of timber" (B.C. Legislative Assembly, 1974: s. 7). Although the legislation did not specify the trading functions enumerated in the Pearse blueprint, Williams envisioned that the proposed board would be an active player in the market. It was to be his cat among the pigeons. As he said several years later:

> *We were setting up a corporate entity to intervene in the market to buy, sell and be a single monopoly exporter of unprocessed raw materials . . . . The idea was: let's explode the scam; we are going to be an honest buyer in this process. If it's available for $50 a cunit —wonderful! We're the buyer. Thanks, we'll take the logs. And we'll resell them for $200 or whatever the real value is. So this was a marvellous way to explode this whole thing . . . and say, "that's just the market operating"* (Williams, 1981: 43-44).

As it turned out, the Forest Products Board was stillborn. The sections of the Timber Products Stabilization Act dealing with chip prices were proclaimed and used to boost the price of wood chips in early 1975, but the broader provisions setting up the board were never proclaimed. The Timber Products Stabilization Act represented the outer limit of Williams' exploration of radical territory. It put in place the legislative authority to implement key elements of his rent collector philosophy. The decision to back away from implementation seems to have been dictated by the deepening effects of the economic down-turn which began in mid-1974 and by the increasing load of political problems on Williams' shoulders. The continued stream of bad news apparently weak-ened his resolve, making it more difficult for him to withstand criticism from the industry and its allies. He retreated to the traditional forests minister's role, increasingly emphasizing the importance of trying to nurse the industry through bad times.

Williams' decisions to back away from plans for the Forest Products Board and stumpage increases did little to assuage the large companies. By 1974 they were in full battle cry, leading a concerted campaign against Williams while working with other corporate players to resuscitate the so-called free enterprise alternative to the NDP.

Williams' various forest company acquisitions played into the opposition parties' hands, providing them with plenty of ammunition for fear mongering about the government's supposed plans to take over the industry. Between March 1973 and February 1974, the NDP government acquired four forest op-erations—Ocean Falls (from Crown Zellerbach), Canadian Cellulose (from Columbia Cellulose and its parent, Celanese Corporation), Plateau Mills (from local owners), and Kootenay Forest Products (from Eddy Match). Although the acquisition decisions were taken independently of one another and dic-tated more by circumstance and pragmatism than by any grand plan, they reflected key Williams' dispositions and strategies. He saw each decision as a way of gaining control over important areas of Crown forest, and thus as a means of ensuring the rent was captured: "my experience was that it was much easier politically to acquire a company than fight a continuing battle to collect the rent" (pp. 87-88). He was hostile to absentee ownership and re-gional concentration. He wanted to acquire means of monitoring the indus-try's dealings from the inside. And by demonstrating that imaginatively-run state enterprises could outperform their private sector counterparts, he hoped to expose what he saw as the industry's dismal record of entrepreneurship. In his words:

> We had inserted ourselves in the game . . . . I think it ended up being a real shocker to the industrial sector that, my god, we really did understand the game that was going on and we had decided to jump in and play it . . . . [I]t must have been terribly disturbing for these people who saw British Columbia just as a corpse to be divided up between a few of them, to have government in playing the same game and maybe playing it better than they were (pp. 27-28).

The takeovers did contribute to some of Williams' goals. They slowed the rate of regional concentration in at least a couple of regions, introduced new actors and ideas, gave Williams and those around him an opportunity to learn about the forest economy from the inside, and put in place important com-ponents of regional development strategies that might have borne fruit had they been pursued after 1975. However, none of this helped to staunch the haemorrhaging of political capital associated with the takeovers. They played neatly into the hands of opposition attempts to portray Williams as the socialist bogeyman bent on wholesale nationalization of the industry. Although Cancel, Plateau and Kootenay Forests Products performed reasonably well after the takeovers, (and Ocean Falls not as badly as many had predicted), the attacks escalated. Throughout the 1974 and 1975 legislative sessions Williams faced a torrent of criticism over operation of the companies, much of it apparently based on information fed to opposition parties by industry insiders.

Forest Land Use Initiatives

Many of the lasting benefits of the 1972 government can be traced back to Williams' decision to establish the Secretariat to the Environment and Land Use Committee of cabinet (ELUCS). Early in the term, Williams rejected a part of the party platform which called for the establishment of a department of Environmental Quality and Planning, indicating that he preferred a coordinative agency to the idea of a super-department veto agency. Composed of three units, ELUCS had a total staff complement of 90 established positions at the end of 1975.

For Williams, ELUCS represented a quick way of implanting a cadre of energetic, loyal, dogma-challenging advisors in the midst of the hidebound bureaucracy he inherited. As someone who relished the problem-solving challenges presented by resource use conflicts, Williams revelled in the hot house climate surrounding the Secretariat. It was an extension of himself, its work a way of implementing his penchant for learning through experimentation. In turn, Williams' strong political support was ELUCS' strongest asset. Because it rode in the bow-wave of the second most powerful figure in the Barrett Government, ELUCS had considerable success as a change agent. It introduced new ideas, broke down inter-agency communication barriers, and prodded calcified parts of the bureaucracy. The Secretariat was not extensively involved in the development of forest policy, however parts of its work had an important indirect impact. It began to implement regional planning structures with the potential to reduce the Forest Service's power. Perhaps more importantly, though, it played a leading role in developing an environmental critique of Forest Service orthodoxy. The negative consequences of past forest management policies began to be documented, both in studies undertaken by ELUCS staff, such as that on northwest development, and in reports commissioned by ELUCS, such as those on the Mica reservoir area and the Purcell range (Wilson, 1987-88: 24; B.C. Environment and Land Use Committee Secretariat; B.C. Environment and Land Use Committee [K.G. Farquharson]; Chambers). These studies played an important early role in deflating the mood of "superabundance", presenting evidence and arguments which supported environmentalists' scepticism about the sustainability of prevailing harvest levels. Each argued that overly optimistic assumptions had resulted in exaggerated timber inventories and over-commitment of the resource to licensees. Harvest levels premised on the inflated inventories would be difficult to sustain. The reports also contained some of the earliest references to the inevitability of timber supply falldown (Wilson, 1987-88: 24).

The Secretariat also played a key part in one of Williams' pet projects, the Burns Lake local economic development experiment. Using the opportunity provided by the issuance of a new forest licence in the area, Williams arranged for the B.C. Association of Non-Status Indians to acquire a stake in a consortium headed by Babine Forest Products. Relying heavily on local leadership, ELUCS coordinated training and other initiatives aimed at reducing unemployment among the community's Native Indian population. The Burns Lake experiment reflected Williams' gravitation towards decentralist approaches:

> *We finally realized that the only answer is to send power out to the regions and get something genuinely diverse and worthwhile happening . . . . You work with the talent that is out there. When there is talent in flower, you grab it, you encourage it, and you give it as much space as you possibly can* (Williams, 1981: 51, 65).

Williams' commitment to experimentation and decentralization was, however, tested when it came time to respond to the powerful manifesto for local control developed in the report of the Slocan Valley Community Forest Management Project (1975). Williams' belief that control should be diversified disposed him to respond positively, but he was wary. As he later said of the Slocan report:

> *I still think it is probably the finest social economic analysis in modern history in British Columbia . . . . There is nothing that comes near it . . . . It was a monumental piece of work. So I was impressed, but I was still a pragmatic politician, saying "how far can we go?" We were talking about the Crown jewels . . .* (Williams, 1981: 52).

His ambivalent response left many Slocan valley backers of the report (and of the NDP) very disenchanted.

The Williams Experience: No Blueprint and Little Capacity

Williams planted some important seeds (and the occasional important land mine). A number of his initiatives contributed significantly to scepticism concerning the liquidation-conversion paradigm that had guided forest policy since the war. A number of NDP measures set in motion, or lent momentum to, processes that led to important integrated resources management reforms in the ensuing years, but the challenge of coordinating the various initiatives soon overwhelmed Williams and the few subordinates he trusted. Faced with what he perceived to be a capital strike (pp. 28-29), and experiencing the pressures associated with the downturn in markets, Williams threw up his hands and conceded the need for a full royal commission. This disappointing performance can to some extent be tied to a lack of capacity. Williams himself was overloaded. At that time, forests was part of a department which also included lands and water resources. Williams was also responsible for B.C. Hydro and, during the first nine months of the NDP's reign, for the Department of Recreation and Conservation. In addition, as one of the ablest ministers in the government, he was frequently called upon to firefight conflagrations breaking out in other departments. This load might have been manageable had he been able to quickly mobilize a full team of sympathetic experts. He was unable to do so. As noted, Williams did get excellent results from Pearse. The Timber Authority idea could have been the nucleus of efforts to diversify the industry and encourage innovative investment. Its fate pointed to a second problem.

Williams' limited exploration of radical terrain can also be linked to his inability (or disinclination) to do what was required to mobilize public support. Several years later Williams provided one diagnosis of the problem:

> *I remember arguing that we had to massage problems in public before we provided the answers . . . . I for one was just too wrapped up in finding new solutions when the public did not even know there was a problem. I was just really enjoying finding these new solutions . . . but the public was back there saying "what the hell is that wild creep doing now?" So the solution would become a problem* (p. 122).

The government's problems in this regard became especially apparent when prices and profits began to slide. Having left himself exposed on so many fronts, Williams was vulnerable to arguments blaming him for all of the industry's problems. In order to limit the damage, he retreated to the traditional forest minister's role of industry booster and protector. As he reflected:

> *". . . it wasn't the same kind of game any more. The money had been flowing in . . . there was a climate to do things. But with the downturn it became necessary to politically align yourself with them and say: hey, poor fellows . . ."* (p. 44).

This explanation does, however, leave the broader question of whether there existed any public appetite for Williams' reform vision. Another interpretation of his record would simply be that he was forced to confront an unpromising potential support coalition. It might have been described as consisting of IWA members who were lukewarm or ambivalent towards the tenure reform/rent collection agenda, environmentalists who were unconvinced of the links between that agenda and their own, small operators who were cowed by contractual connections to the majors, and local control supporters who wanted more or were unable to amplify their support. Stated differently, we could say that Williams came face to face with a stunted public imagination, with the fact that British Columbians had never been encouraged to adopt and assert the landlord's perspective. A look at the experiences of the NDP government elected two decades later will give us an opportunity to reflect on whether these basic realities have changed.

PLANNERS AND LAWYERS VS. RADICAL REFORMERS: THE HARCOURT GOVERNMENT EMBRACES A MODERATE REFORM AGENDA

The Harcourt Government implemented an extensive package of moderate forest policy reforms. Its initiatives required the mobilization and direction of an immense amount of bureaucratic and political energy. The government deserves marks for having pulled this off in an era when state authority and capacity is everywhere in decline, and at a time when media induced panic seems to be almost a necessary condition for the "accomplishment" of any major public policy end.

The breadth and relative coherence of the Harcourt package are in considerable part a result of the far-reaching policy development work done by the bureaucracy (and, more generally, by the policy community) in the years prior to the NDP's 1991 victory. Whereas Williams found the blueprint drawer empty, Harcourt's first Minister of Forests, Dan Miller, found an extensive menu of reform designs.

In selecting from this menu, Miller and his successor, Andrew Petter, avoided the tenure issue, choosing an alternative course towards their goals of "peace in the woods", and a soft landing transition to a more sustainable forest economy. The moderate initiatives chosen were supported by the IWA and very appealing to the cadre of environmental lawyers and planners whose fortunes rose with the arrival of the new government.

The Harcourt reform package rests on four cornerstones: the Commission on Resources and Environment (CORE) and the Protected Areas Strategy (PAS), the timber supply review and annual allowable cut (AAC) redetermination process, the Forest Practices Code, and the Forest Renewal plan. As we will see, the first three of these were based on policy thrusts initiated in the bureaucracy prior to the NDP takeover. The new government's endorsement of these three initiatives locked it into a search for the fourth, and ruled out any thoughts of tackling the tenure issue.

84   *Troubles in the Rainforest*

The Antecedents of the NDP Reforms

The well developed reform agenda inherited by the Harcourt Government reflected the increased assertiveness of the natural resources bureaucracy during the final years of Social Credit. Capitalizing on the atmosphere of political drift that set in as Social Credit tried unsuccessfully to deal with the Vander Zalm scandals and the transition to Rita Johnston, the forests, environment, and parks bureaucracies all took stock and began to craft plans. Most importantly, the Ministry of Forests (MOF) bureaucracy capitalized on the pro-change mood that took hold in cabinet after Forests Minister Dave Parker was pummelled during province-wide public hearings on his July 1988 proposal to encourage the roll-over of FLs (volume-based tenures) into TFLs (area-based tenures). Pressure for change was reinforced by a number of reports chronicling poor industry and ministry performance in the wake of the 1982 adoption of "sympathetic administration" policy.[1] As it had in response to the first wave of IRM pressure in the 1970s, the MOF showed its adaptive resourcefulness. Correctly anticipating that the major manifestation of pro-change mood, the Forest Resources Commission (appointed in 1989), would make recommendations with some negative implications for its status, the ministry began crafting the key parts of an alternative package of moderate reform proposals which would enhance or maintain its role.

We will now consider the antecedents of each of the cornerstone initiatives. The forces which led to CORE and the PAS had considerable momentum by the time the new government took over. Two antecedents were crucial. First, in the wake of the 1986 report of the Wilderness Advisory Committee (WAC), various government and public groups and agencies began to design proposals for a more comprehensive approach to land use planning. Second, in the late 1980s both the Forests and Parks bureaucracies embarked on processes aimed at developing procedures for comprehensive review of candidate areas for the protected areas system.

The WAC was presented with a number of persuasive briefs on the need for a comprehensive approach to the evaluation of park-wilderness candidate areas and to provincial land use planning in general.[2] In the years after 1986, the need for a comprehensive land use strategy was a dominant theme in the deliberations of the Dunsmuir I and II gatherings (1988 and 1991, respectively), the Forest Resources Commission (FRC), the B.C. Round Table on the Environment and the Economy, the Old Growth Strategy process, the Parks and Wilderness for the 90s process, and the B.C. Forest Industry Land Use Task Force.[3] Most of these recommendations emphasized the potential of consensus building approaches. In this and other respects, developments during this period were strongly influenced by moderate environmentalists from organizations like the Outdoor Recreation Council (ORC) and by sympathetic bureaucrats. In 1991, a number of these bureaucrats came together in the "PLUS" (Provincial Land Use Strategy) working group. While these musings about process unfolded, both the Parks and Forests bureaucracies began to design more direct responses to concerns about the preservation of wilderness and old growth. Key elements in the MOF response included the Old Growth Strategy Project, and a process aimed at identifying and designating wilderness areas. The Parks Branch's new assertiveness was reflected in Parks Plan 90, a comprehensive planning process aimed at identifying candidate areas. In December 1990, the Forests and Parks protected area initiatives were pulled together in an exercise called "Parks and Wilderness for the 90s". In early 1991, the two ministries held joint open houses across the province to solicit public input on their lists of proposed protected area candidates.

While it is unlikely that the institutional landscape would have evolved exactly as it did had the NDP not been elected, CORE and the PAS were fairly logical consequences of the design work done in the bureaucracy and the policy community between 1986 and 1991. The critical element added by the new government was, of course, its commitment to double the proportion of the province under protected area designations to 12 percent by the year 2000.

Taking their cue from this pledge and from Harcourt's campaign commitment to end "valley-by-valley" confrontations, and building on the work of the PLUS group and others, senior bureaucrats from forests, environment, and the Cabinet Office worked quickly to design the Commission on Resources and Environment (CORE). Harcourt unveiled CORE and its Commissioner, Stephen Owen, in January 1992, noting that it would lead the process of developing a comprehensive land use strategy, and that it would undertake a regional planning process on Vancouver Island during the next 18 months. By the time legislation establishing its mandate was passed in July 1992, CORE had been asked to initiate additional regional planning processes covering the Cariboo-Chilcotin, and the Kootenay-Boundary (later split into the East Kootenay and the West Kootenay-Boundary). Its proposed land use plans for these four regions were submitted between February 1994 and October 1994. In the meantime, CORE began to outline the elements of a land use strategy. These included a Land Use Charter, accepted in principle by the government in June 1993, and a long list of provincial land use goals (B.C. Commission on Resources and Environment, 1993, 1994). In late 1994 and early 1995, it released recommendations concerning a sustainability act, and an improved planning delivery system (B.C. Commission on Resources and Environment, 1994-95).

The Protected Areas Strategy (PAS) was announced in May 1992. It was designed to carry forward the work of Parks and Wilderness for the 90s and the protected areas part of the Old Growth Strategy Project. The various parts of the bureaucracy involved were handed a timetable for evaluating a list of 184 study areas. This list, which was substantially the one developed by Parks and Wilderness for the 90s, was augmented in August 1992 by the addition of 14 areas which had been recommended for deferral by the Old-Growth Strategy Project. Decisions resulting from the PAS process were to be factored into the broader land use planning initiatives being carried out through CORE regional processes or sub-regional land and resource management planning (LRMP) processes. By late 1994, the four CORE regional processes and the 12 LRMPs under way covered over 50 percent of the province (CORE, 1994-95, Volume 1, 31-32).

The NDP's attempts to come to grips with the inevitability of timber supply reductions had their immediate antecedents in a 1990-91 Forest Service internal review of its timber supply analysis and harvest level determination procedures (B.C. Ministry of Forests, March 1991 and April 1991). The review team's report was a clear admission of the ministry's inadequacies. It pointed out that the budget cuts and downsizing of the 1980s had left the ministry incapable of performing the most basic of forest management functions—the determination of the allowable annual cut (AAC). According to the report:

> *There is a perception among many staff that allowable annual cuts are too high . . . . Timber Supply Area timber supply analyses are not being performed rapidly enough to stay current with changes in management procedures and policies . . . . Many staff stated that current AAC levels do not allow for the delivery of sound integrated resource management . . . . The Forest Service has lost its leadership in the field of timber supply analysis. This is due, in part, to giving this function a low priority; the Forest Service has not aggressively recruited and maintained its own expertise* (B.C. Ministry of Forests, March 1991: 4-5).

The corrective plan, released in April 1991, called for closer integration of timber supply planning and forest land use planning, a review of sustained yield policy, clearer definition of roles and responsibilities of those responsible for TS planning, and establishment of a Timber Supply unit at MOF headquarters (B.C. Ministry of Forests, April 1991). The report also recommended a three year timetable for completion of AAC reviews in all TSAs. These reviews were to "bring AACs up to date with current management practices" (p. 12).

Miller endorsed this plan. By mid-1992 he pushed through Forest Act amendments establishing December 31, 1995 as the deadline for the determination of new AACs for all TSAs and TFLs. The Forest Service launched into a three year timber supply review process.

Prior to the election the NDP pledged to develop a new forest practices code. Its new law would, among other things, control the size of clearcuts, encourage selective logging on appropriate sites, preserve fishery values and water quality, minimize road building, and preserve the maximum possible biological diversity. While in opposition, Miller presented a Private Member's Bill aimed at ensuring sound forest practices on private land.

These campaign pledges converged neatly with development work done in the bureaucracy and the FRC prior to the election. In its April 1991 report,

The Future of our Forests, the FRC recommended that the disparate rules and standards found in various statutes should be amalgamated into one all-encompassing, stand-alone Forest Practices Act (B.C. Forest Resources Commission, 1991: 87-88). In July 1991, Minister of Forests Claude Richmond released a discussion paper on a code (B.C. Ministry of Forests, July 1991). It listed the advantages and disadvantages of several implementation options, including a voluntary code, a self-administered code, a contractual code, and a legislated code.

There was little question which option Miller would favour. The idea of putting in place a legislative basis for firm regulation of forest practices was supported by the IWA, the forests bureaucracy, and the environmental lawyers and planners who wielded so much influence in the forest policy community. From the outset, the idea of a code was embraced by the new government as a means of demonstrating its commitment to better management. After the election, the idea of a code quickly became a cornerstone of its overseas public relations campaign aimed at countering boycott pressures.

In January 1992 Miller asked the FRC to make recommendations concerning the legislative and administrative framework for a code (B.C. Forest Resources Commission, July 1992). In May 1992 he handed responsibility for technical development of a code to a steering committee composed of bureaucrats from Forests and other ministries. The development process moved quickly. By mid-1994, the government had introduced and passed the 132 page Forest Practices Code of British Columbia Act; drafted a number of regulations for public review; released three reports summarizing input received on a "Forest Practices Code Discussion Paper and Rules" (distributed in November 1993); issued a detailed compilation of rules and standards (the "Forest Practices Code Standards with Revised Rules and Field Guide References"); commenced organizational changes in district and regional MOF offices including establishment of Compliance and Enforcement Teams; commissioned and received three independent assessments of the costs and benefits of the code; begun developing the several dozen field guides needed to support standards and regulations; and established training programmes and other elements of the implementation scheme.[4]

It was apparent from the outset that while the code, land use, and timber supply review initiatives might earn marks for promoting sustainability and helping to counteract boycott threats, a reform platform built from these planks would collapse unless it was tied to measures that would ameliorate the transition problems faced by forest workers and forestry dependent communities. Certainly no-one in the government was prepared to contradict IWA assertions about the sanctity of forest workers' jobs. The Premier's repeated assurances on this score locked the government in to a search for a way of socializing the costs of forest industry transformation.

The key institutional actor in this search was the Forest Sector Strategy Committee (FSSC). Its inception could be traced to the Premier's expressed desire to develop strategies (or "sectoral agreements") for key economic sectors, and more directly to a series of late 1992 meetings between government officials and a coalition of forest industry and forest labour representatives who had begun to discuss shared concerns over allowable cut reductions and erosion of the working forest.

The composition of the committee was announced by the Premier in April 1993. It included nine company executives, leaders of three industry organizations and three unions, the Mayor of Prince George, a First Nations representative, the Dean of the University of British Columbia Forestry School, a representative of Forestry Canada, and a Professor of Ecology (later replaced as de facto environmental representative by a forestry consultant highly respected in the environmental community.)

The FSSC's major accomplishment is the Forest Renewal Plan. It was put together through negotiations in and around the committee, with the Premier's top bureaucrat, Doug McArthur, playing a lead role, along with Petter. The plan was announced at a well-publicized launch on April 14, 1994. A new Crown agency, Forest Renewal B.C., was to be established to allocate forest renewal investments. According to the minister, it would be guaranteed "a dedicated and specific source of revenue" from stumpage increases (B.C. Legislative Assembly, 1994: 10225). These increases were predicted to bring in an average of about $600 million extra per year over the period to 1999. About $400 million per year would flow into the Forest Renewal pot. These, the Minister stressed, would be new dollars, over and above current budgets for silviculture and forest management (B.C. Legislative Assembly, 1994: 10384)

Approximately half of Plan investments would go to reforestation and silviculture, to reclamation of marginal agricultural land and other measures to increase lands available for tree planting, and to research and development. According to the glossy document presented at the launch, these measures would offset an anticipated 15-30 percent reduction in cut levels over the next 50 years. The new agency's value-added priority would be pursued through investments in value-added companies and research, in forest worker training, and in community economic development and diversification. The announcement of the forest renewal deal naturally lead to questions of how industry had been induced to swallow the huge stumpage increase. There was persistent speculation that the agreement had been underwritten by "understandings" that the governments would go slow on tenure reform and AAC reductions.[5]

Whatever the nature of the deal, it bears noting that in order to accomplish it, the government had to adopt a policy style very different from those used in implementing other parts of its agenda. While some of those processes might be described as hyper-consultative, the path to the forest renewal agreement bore all the marks of corporatist, "iron triangle" deal making.

The Path Not Taken: Putting Off Tenure Reform

Without any apparent systematic analysis of the idea, the NDP spurned an alternative path premised on the view that major tenure reform would be the most effective route to the goals of sustainability and industry transition. It also avoided structural reforms of the sort embodied in calls for devolution of forest management authority to community forest boards.

Why did the government avoid the tenure-structural reform course? On one level at least, an explanation can be constructed from obvious elements. Before we turn to that explanation, it needs to be stressed that the choice was not because of a lack of blueprints for tenure reform and structural change. In the years immediately prior to its election, key NDP actors had no doubt familiarized themselves with the decentralist, community forest board proposals sketched by Herb Hammond, Michael M'Gonigle, the Village of Hazelton, the Tin-Wis Coalition, and others (Hammond, 1991; M'Gonigle, 1986; M'Gonigle, 1989-90; Tin-Wis Coalition, 1991; Village of Hazelton, 1991; Taylor and Wilson, 1994). And members of the new government knew well the detailed tenure reform proposals presented by the Forest Resources Commission (FRC) and the Truck Loggers Association (TLA).

The FRC sought to design a way to reduce the amount of the cut allocated under secure licence arrangements from about 85 percent to 50 percent. It held out the promise of greater tenure security in exchange. The cut thus freed up would be used to diversify tenures, with some of it dispensed under new volume-based tenures, and some reallocated to small area-based tenures managed by communities, Native Bands, and woodlot operators. These small license holders would feed a competitive log market. With their guaranteed supplies reduced to approximately 50 percent of their needs, companies with manufacturing facilities would turn to this market to meet their remaining requirements (B.C., Forest Resources Commission, 1991: chap. 5).

The FRC said that by helping ensure realization of the full value of the resource, the log market would generate the revenues required to fund enhanced stewardship. After noting that since the turn of the century at least 14 permanent silviculture funds had been set up and abandoned, the FRC argued that a Crown corporation represented "the strongest mechanism available to secure stable, long-term funding for enhanced forest stewardship" (p. 28). It recommended establishment of a Forest Resources Corporation. This corporation would collect revenue from designated forest lands, using the money raised to renew forests.

The FRC drew heavily on advice from the Truck Loggers Association. In its submissions, the TLA argued that "the tenure system in B.C. has become closed, rigid and—in terms of LRSY trends, silvicultural performance, and growth and yield standards—sterile and inefficient" (Truck Loggers Association, 1990b: 29). The TLA's initial brief called for complete separation of forestry and manufacturing operations, with processors required to obtain timber through a timber market. The entire working forest would be allocated to area-based tenures. A diverse range of tenure forms was proposed, with a maximum management unit size of 200,000 hectares recommended (Truck Loggers Association, 1990: 19-21). The timber supplies of existing conversion plants would be guaranteed at a "reasonable price" during a five year transition to market prices (p. 19).

After extensive consultation the TLA moderated its position. Bowing to criticism that its initial transition guarantees were insufficient, it sought to reassure existing operators by proposing that existing operations would hold onto about half of their existing apportionment in the form of new area-based tenures linked to processing facilities. These licences would account for about 40 percent of the cut. Another 40 percent would be apportioned to new, smaller-scale area-based tenures called forest management tenures, with the remaining 20 percent dedicated to woodlot licences. Harvest from the latter two tenures would be sold on the open log market.

The TLA believed that the inability of the provincial government to properly fund silviculture made necessary a new approach to management and forest financing. Accordingly, it said, responsibilities should be turned over to a new Crown-led company, dubbed ForestCo. It would be assigned the rights to, and management responsibility for, the designated working forest. Its primary purposes would be forest renewal and the attainment of the highest possible growth and yield standards. It would invest in silviculture directly and through advances to the diverse tenure holders, collecting returns through stumpage and a share of log market sales (Truck Loggers Association, 1990b: 34-37).

Shortly after taking over, the NDP was also forcefully reminded of some of Peter Pearse's 1976 recommendations on tenure reform. The reprise came in the report of the Commission set up to inquire into the issue of compensation for resource tenures "taken" by the Crown for other purposes. In the strongest part of his report, Commissioner Richard Schwindt recommended giving the government a much freer hand to withdraw AAC for non-timber uses (B.C., Resources Compensation Commission, 1992: 110-114). The current legislation allows a non-compensable 5 percent takeback every 25 years from TFLs and a non-compensable 5 percent takeback every 15 years from Forest Licences. Schwindt endorsed the 1976 Pearse Royal Commission proposals that the legislation governing both TFLs and Forest Licences should allow for a non-compensable 10 percent takeback every 5 years. In making this recommendation Schwindt noted that "reductions attributable to re-inventorying, changed harvesting standards and similar factors are not to be included in this ceiling" (p. 114).

The industry's hostile reaction to the Schwindt proposals was no doubt conditioned by the residual anger it felt over the 5 percent reduction in quota imposed by the Social Credit Government in 1988 in order to boost the timber allotment to participants in the Small Business Forest Enterprise Program. Smarting over having been forced by a friendly government to swallow what it perceived as a dose of confiscationism, the industry was determined to discourage any parallel designs that might lurk in the minds of the NDP. This anger clearly got through to the government. When he released Schwindt's report for discussion Attorney General Gableman was at pains to reassure the industry that the government had already buried this part of the report.

Why was the Harcourt Government so averse to addressing tenure reform and fundamental structural change? The risk aversive course chosen was completely in keeping with the pre-election emphasis on inoffensiveness. No-one in the government (particularly those who had watched the majors' assault on Williams) believed that it should risk becoming bogged down in a nasty and prolonged battle with the industry. The party had not sought a mandate for tenure reform; there is no way such a campaign plank would have been acceptable given the Harcourt party's emphasis on moderation. No-one in Cabinet believed tenure reform could be accomplished without draining political capital; while it was recognized that there was considerable public support for the general concept of tenure reform, there were grave doubts about the depth and resiliency of the support for any particular version. Clear signs of the limitations of the potential support coalition had been apparent when the FRC proposals sank like a lead balloon in 1991. The commission's lack of impact was no doubt partially due to its confusing presentation. But more broadly, its failure to galvanize and unite the forces who paid lip service to the need for tenure reform sent a clear message about the political risks associated with the issue.

It is also worth noting that neither Miller nor Harcourt was at all sympathetic to tenure reform. Even if they had been, they might well have been dissuaded by the depressed condition of the industry at the time they took over. In down market times there are strong pressures on any minister to concentrate on helping the industry recover from the last red ink period. Petter was more convinced of the need for tenure reform, and by the time he took over in 1993 the industry's profit picture was perhaps more conducive to thoughts of reform. But by this point, the first term agenda was set. It had been accepted that in order to keep the sustainability initiatives afloat, the industry had to be coaxed into the forest renewal deal. The industry presumably did not have to spell out its position that such a deal would be jeopardized by opening up the tenure issue. With little choice, Petter became a quite persuasive advocate of the view that the government's main policy thrusts represented logically prior steps to a consideration of tenure:

> . . . *to move on tenure reform before you have a clear sense of what your land use allocation is, and what your forest management regime will be, and what the extent of your resource is, is to engage in a mug's game. How can you possibly determine how you are going*

> *to apportion harvesting rights across a land base until you first*
> *determine what that land base is, what the extent of the resource is,*
> *and how you are going to manage that resource? Once you've done*
> *that then it makes sense to address what I agree is a very important*
> *part of the equation and a part that cannot be left out of the equa-*
> *tion if we are to make the full shift* (Petter, 1995).

The tenure reform forces have a lot more work to do. Despite two decades of very persuasive advocacy from the likes of Drushka and M'Gonigle, there remain doubts about what kind of a support coalition can be mobilized (Drushka, 1985; M'Gonigle and Parfitt, 1994). It is not clear that an objective reading would be very different than our assessment for the 1970s. In order to achieve the necessary broadening and deepening of public support, advocates of tenure reform must address the IWA's wariness, convince urban environmentalists of the relevance of the issue, bridge the gulf between reform sympathizers on either side of the private ownership issue, and present a transition plan a lot more convincing and reassuring than the ones outlined by the FRC and the TLA. Thus, on one level, explaining the Harcourt Government's reluctance to engage issues of tenure reform and structural change simply entails stating some obvious points about political feasibility given prevailing values and power structures. Obviously, however, this level of analysis avoids more difficult questions about obstacles to the kind of "paradigm shift" cultural and political transformations required to change those values and structures.

COMPARING THE TWO GOVERNMENTS

The results of the comparison underline a couple of obvious points about forces that promote or impede policy change. First, reform- minded governments are constrained by their inheritance, particularly by the policy ideas and blueprints "in the mill" when they take over. Second, in economic policy making areas like the one examined here, a government's change agenda—and its success in implementing it—will be influenced by fluctuations in the health of the regulated industry.

Forest policy making is an exercise in grappling with complexity and uncertainty. New governments wishing to accomplish change face the challenge of transforming this uncertainty and complexity into a change design sufficiently compelling to allow for the mobilization of political and bureaucratic resources. Since our parties have weak policy development capacities and generally prefer to present the electorate with vague policy proposals, blueprints of sufficient clarity and coherence are more likely to generate from the bureaucracy or the policy community than the party. The quality of blueprints inherited by a new government are critical, particularly for one wishing to move quickly.

The two governments analysed here worked with radically different policy idea inheritances. Finding no blueprints for translating his critique and vision into policy, and perceiving a paucity of sympathetic expertise, Williams had little alternative but to scramble. He tried to combine pro-active and reactive approaches. He initiated some policy design work, but more importantly, he tried to capitalize on opportunities that presented themselves to sow some seeds. He hoped these would contribute to the desired sorts of changes. Williams' cleverness, energy, and penchant for experimentation equipped him well to lead such an approach. Much was accomplished. Nonetheless, with so many things "on the go", control and coordination problems soon began to escalate. By the end of the third year, Williams was ready to accept the need for a comprehensive royal commission examination of core forest policy issues.

The Harcourt Government, on the other hand, found a full blueprint drawer. Its experience illustrates a good recipe for quick implementation of a reform agenda in a complex policy area: start with a bureaucracy committed to well-developed reform plans; add one fresh government with a general mandate to push those sorts of reforms. By throwing their weight behind initiatives that were thoroughly developed and, in a couple of cases, already into the implementation phase, Miller and his colleagues were able to make a fast and fairly coherent start on a reform agenda.

It might, of course, be argued that in choosing the course it did, the Harcourt Government allowed the civil service to dictate its plans. As we have noted, the legacy of ideas passed to the NDP in 1991 included some blueprints for different and more radical reform paths. No doubt the bureaucracy's lack of sympathy was a determinant of their rejection. We have argued, however, that a variety of other factors influenced the rejection of the tenure and structural reform option. There was an obvious fit between the plans the bureaucracy had under development and the Harcourt Government's commitment to a moderate agenda.

Turning to a second point of difference, the two governments were affected very differently by forces connected to ups and downs in industry profits. The forests minister (and the ministry) must perform a variety of roles. A simplistic distinction can be drawn between the regulator-reformer role, which gives priority to protecting the Crown/public interest in the forests, and the industry booster role, which gives priority to protecting the industry and its workers. Forests ministers are constantly balancing those roles, searching for compromises that seem appropriate to the times and likely to conserve political capital. Where ministers position themselves will obviously depend on their ideological inclinations. As well though, this choice will depend on the economic condition of the industry and on the external factors shaping that condition. When the industry is in a down cycle ministers are subjected to powerful pressures to adopt the booster/protector stance, and to blur the distinction between the industry interest and the public interest. The chances of policy reform are greater during profitable periods, although it can be argued that those chances are still somewhat minimal. On the up cycle, forests ministers will certainly hear arguments about the importance of giving the industry an opportunity to "get its balance sheets back in order" after the last recession.

When it took over in late 1991, the NDP inherited a rather gloomy situation. The forest industry was into the second year of a downturn; it had posted losses in 1990, and was on its way to much bigger losses in 1991. Worries about consumer boycotts had become a major factor as a result of negative media coverage in Europe and the USA. And, in response to the September 1991 cancellation of the 1986 MOU, the U.S. Coalition for Fair Lumber Imports was gearing up for its third trade assault in a decade.

As a result of these factors Miller was under immediate pressure to give primacy to the booster role, a stance to which he was favourably disposed as a result of his sympathies for forest workers. Under the circumstances it is not surprising that Miller and Harcourt showed absolutely no appetite for the radical portion of the reform menu. What should perhaps be treated as surprising is that the new government sampled as extensively as it did from the moderate side.

As we saw, the upturn in the market and profit cycle which began in 1993 did not translate into a shift towards more radical agenda. Given its commitment to finding the deal needed to keep the sustainability initiatives afloat, the government did not want to jeopardize the oncoming flow of revenue. The industry upturn was a gift from the gods that opened the opportunity to develop the politically crucial missing part of its programme. With revenues flowing, the government was able to move towards stumpage increases and the forest renewal plan.

Williams faced the mirror image of this cycle. After experiencing buoyant markets for the first 18 months of the Barrett regime, the industry turned down in mid-1974. As we saw, this brought a multi-dimensional compounding of Williams' problems. His stumpage-royalty increase moves were derailed, and the Forest Products Board plan was dropped. Soon he succumbed to pressure to adopt the traditional booster/protector role.

Thirdly, the two governments differed in their emphasis on, and aptitude for, the challenges involved in mobilizing the public support for change. As Williams acknowledged, he neglected this element of the enterprise. The Harcourt government, on the other hand, put considerable emphasis on the need to galvanize support and develop consensus. It was quite adept in this regard, utilizing approaches ranging from the hyper-consultative "stakeholder" processes involved in developing CORE's regional plans, to the traditional linear public participation exercises used in developing the Forest Practices Code and in preparing for allowable cut decisions, to the corporatist, deal-making style employed to strike the Forest Renewal deal. As expected, the comparison tells us more about changes in the contours and boundaries of this policy sector than about enduring determinants of government inclinations and effectiveness. The two governments spanned a period of remarkable change in forest policy sector. Over the course of two decades the province shifted from a mood of superabundance to a mood of limits, from the easy politics of allocating an expanding pie among a limited list of adversaries who shared many key values, to the much more difficult politics of allocating a shrinking pie among a longer list of adversaries who were divided by value differences. The difficulties of the new politics are compounded by the fact that many of the issues engage intense emotions of identity and belonging. Some of the combatants are on moral-symbolic crusades, while others are desperately committed to maintaining full-bore operation of logging and manufacturing capacity to which they link their jobs and/or ability to make bank payments.

The transition to shrinking pie politics gets entangled with two other interrelated shifts in the character of the policy field. First, there was a shift towards a more participative, consultative approach to policy making. We have moved in the course of 25 years from an era when the public was for all intents and purposes frozen out of deliberations on forest and land use policy, to the present situation which each year finds thousands of British Columbians devoting perhaps hundreds of thousands of hours of volunteer time to various facets of the forest policy process. And second, we see a coincident transformation in the nature of the knowledge foundation underlying forest policy. In an attempt to highlight the extent to which the assumptions and knowledge claims underlying forest policy have become both more transparent and more contested, we might awkwardly characterize this as a shift from a dogma-based foundation to a science-based foundation.

We are only beginning to reflect upon the implications of these shifts. There are implications in terms of how policy is made and legitimated, in terms of what goes on at the interface between science and politics, in terms of how uncertainty is used and responded to, and in terms of the nature of our democracy. All of these need to be explored.

ENDNOTES

1 See, for example, B.C., Office of the Ombudsman (1991); Nixon (1987), Nixon (1989).

2 See, for example, submissions to the WAC by Stephen Fuller, Ian McTaggert-Cowan, the Outdoor Recreation Council, and the Federation of Mountain Clubs of B.C., along with the background paper prepared for the WAC by Murray Rankin.

3 For brief reviews of these, see B.C., Commission on Resources and Environment (1994), The Provincial Land Use Strategy, Volume 2, Appendix 5).

4 See B.C., Ministry of Forests, "Forest Practices Code Update." Numbers 2-10 (February 10, 1993 -January 1995).

5 See, for example, Gordon Wilson in B.C., Legislative Assembly (1994). Debates, April 25, 10228-9.

REFERENCES

B.C. (1994). British Columbia's Forest Renewal Plan. Victoria.

B.C., Commission on Resources and Environment (1993). 1992-93 Annual Report to the Legislative Assembly. Victoria.

_____ (1994). 1993-94 Annual Report to the Legislative Assembly. Victoria.

_____ (1994-95). The Provincial Land Use Strategy (4 Volumes). Victoria.

B.C., Environment and Land Use Committee (K.G. Farquharson) (1974). Final Report: Mica Reservoir Region Resource Study. Victoria.

B.C., Environment and Land Use Committee Secretariat (1976). Terrace-Hazelton Regional Forest Resources Study. Victoria.

B.C., Forest Resources Commission (1991). *The Future of Our Forests*. Victoria: Forest Resources Commission.

_____ (July 1992). Providing the Framework: A Forest Practices Code. Victoria: Forest Resources Commission.

B.C., Legislative Assembly (1974). Statutes. Timber Products Stabilization Act.

_____ (1994). Debates. April 25 and April 28.

B.C., Ministry of Forests (March 1991). Review of the Timber Supply Analysis Process for B.C. Timber supply Areas: Final Report. Victoria.

_____ (April 1991). Proposed Action Plan for the Implementation of Recommendations from the Report: "Review of the Timber supply Analysis Process". Victoria.

_____ (July 1991). A Forest Practices Code: A Public Discussion Paper.

B.C., Office of the Ombudsman (Feb. 12, 1991). "Letter to Jim Gillespie re: Nahmint case." Excerpted in "The Nahmint Valley: A Case History of Forest Planning Abuse." Forest Planning Canada 8(1), 12- 19.

B.C., Resources Compensation Commission (1992). Report of the Commission of Inquiry into Compensation for the Taking of Resource Interests.

B.C., Task Force on Crown Timber Disposal (February 1974). Crown Charges for Early Timber Rights. Victoria.

_____ (July 1974). Timber Appraisal. Victoria.

Chambers, A.D. (Study Coordinator) (1974). *Purcell Range study: Integrated resource management for British Columbia's Purcell Mountains*. Vancouver.

Drushka, K. (1985). *Stumped: The forest industry in transition*. Vancouver: Douglas and McIntyre.

_____ (1987). B.C.'s Forests: Condition Critical. *The Truck Logger*, Dec./Jan.: 11.

_____ (1990). The new forestry: A middle ground in the debate over the region's forests? *The New Pacific*, 4: 7-24.

Hammond, H. (1991). *Seeing the forest among the trees—The case for wholistic forest use*. Vancouver: Polestar Press Ltd.

M'Gonigle, M. (1986). From the ground up: Lessons from the Stein River Valley. In W. Magnusson, et al. (eds.), *After Bennett: A new politics for British Columbia* (pp. 169-191). Vancouver: New Star Books.

_____ (1989-90). Developing sustainability: A native/environmentalist prescription for third-level government. *B.C. Studies*, 84: 65-99.

M'Gonigle, M., and Parfitt, B. (1994). *Forestopia: A practical guide to the new forest economy*. Madeira Park, B.C.: Harbour Publishing.

Nixon, B. (1987). Promises without substance: The Gairns Report on British Columbia's Nass Valley. *Forest Planning Canada*, 3(6): 15-19.

_____ (1989). Executive management failures in the forest service: The story behind the Doman Industries conflict. *Forest Planning Canada*, 5(6): 41-46.

Pearse, P.H. (1976). Timber rights and forest policy in British Columbia. Report of the Royal Commission on Forest Resources (2 vols.). Victoria: Queen's Printer.

Petter, A. (1995). Interview. (January 18, 1995).

Slocan Valley Community Forest Management Project (1975). Final Report.

Taylor, D., and Wilson, J. (1994). Ending the watershed battles: B.C. forest communities seek peace through local control. *Environments*, 22(3), 93-102.

Tin Wis Coalition (1991). Community Control, Developing Sustainability, Social Solidarity. Vancouver: Tin Wis Coalition.

Truck Loggers Association (1990). *B.C. forests—A vision for tomorrow (an overview)*. Vancouver: TLA.

_____ (1990a). *B.C. forests—A vision for tomorrow (working papers)*. Vancouver: TLA.

_____ (1990b). *Options for the Forest Resources Commission: Review, reconsideration, recommendations*. Vancouver: TLA.

Village of Hazelton (1991). *Framework for Watershed Management (formerly the Forest Industry Charter of Rights)*. Hazelton, B.C.: The Corporation of the Village of Hazelton.

Williams, B. (1971). British Columbia timber. Ripping off B.C.'s forests. *Canadian Dimension*, 7(7): 19-21.

Williams, B. (1971a). Background and point paper, natural resources. New Democratic Party Provincial Convention Papers, 1971.

Williams, B. (1981). Interview. Conducted June 25 and June 30 with Norman Ruff and Jeremy Wilson for the B.C. Project.

Wilson, J. (1987-88). Forest conservation in British Columbia, 1935-85: Reflections on a barren political debate. *BC Studies*, 76: 3-32.

_____ (1990). Wilderness politics in B.C.: The business dominated state and the containment of environmentalism. In W.D. Coleman and G. Skogstad (eds.), *Policy communities and public policy in Canada*. Mississauga: Copp Clark.

Plate 10 (overleaf)   Loading for export ➤

Colonial Vestiges: Representing Forest Landscapes on Canada's West Coast*

Bruce Willems-Braun

Department of Geography, University of California at Berkeley

* This essay first appeared in *BC Studies*, Winter 1996-1997

INTRODUCTION: COLONIAL LEGACIES

> *What 'post-colonial' certainly is not is one of those periodizations based on epochal 'stages', when everything is reversed at the same moment, all the old relations disappear for ever and entirely new ones come to replace them. Clearly the disengagement from the colonizing process has been a long, drawn-out and differentiated affair* (Hall, 1996: 247).

Speaking before the Pearse Commission in 1975, Simon Lucas, then chair of the West Coast District Council of Indian Chiefs, outlined his people's frustrations: 'We feel more isolated from the resources to which we have claim than at any time in the past,' to which he added simply, 'this is becoming more so.'[1]

This poignant statement articulated the experience of many Natives during and after the rapid expansion and consolidation of British Columbia's coastal forest industry in the 1950s and 1960s. In the 20 years preceding the Commission, traditional "Nootka" territories on the west coast of Vancouver Island had been rationalized within the regional and global space-economies of industrial capitalism and in the interests of investors and forestry-based communities far removed from their villages. Segregated on reserves, surrounded by forests of immeasurable value, the Nootka increasingly found that they had almost no access to the resource wealth of their lands, except on the shifting margins of the white wage labour force. Today, almost 20 years later, it is not clear that much has changed. As evident in submissions to a recent Task Force on Native Forestry (BC, 1991), many Native groups continue to find themselves marginalized in their own territories: as workers they remain under-represented in an industry where seniority and Native seasonal practices have been at odds; few hold forest tenures; and, as was evident yet again in events leading to the Clayoquot Sound land-use plan of 1993, decision-making often occurs from a distance—spatial and institutional.[2] The result has been that Native communities have seen the landscapes and ecologies historically tied to their cultures and polities increasingly remade by industrial forestry.[3]

This marginalization has not gone unchallenged, as non-Natives in British Columbia have been made forcefully aware. From the Nuu-chah-nulth in Clayoquot Sound to the Haida in Haida-Gwaii and the Gitksan and Nisga'a in the Skeena and Nass valleys, Native peoples have fought to be heard amid the often rancorous debates in non-Native society over the fate of the forests and forestry in British Columbia.[4] Yet, as the Task Force discovered, Native voices—if heard at all—are usually incorporated as one among many 'special interests' in systems of forest management that were founded upon earlier colonial divisions of space that separated and segregated Native 'reserves' from 'crown lands'. Today, the marginalization of Native peoples on their own traditional territories often appears 'common-sense', in part because the power-relations generated by colonialism have become buried—and thus reproduced daily—in the categories and images through which disputes over the environment, resources, and economic development are framed.

This chapter explores the *discursive technologies* by which First Nations continue to be excluded from discussions of 'rights of access' and 'responsibilities of use' in British Columbia's forests by interrogating a series of *representational practices* which in different ways have worked to abstract the 'forest' from its cultural 'surrounds' and re-situate it within very different cultural logics—the 'market', the 'nation' and, recently, the global 'biosphere'. The significance of these practices, this chapter argues, is that they authorize very different people to 'speak for' the forest: no longer traditional 'owners' and 'stewards' (such as articulated in the Nuu-chah-nulth system of *ḥaḥuulhi*), but rather forestry corporations, professional foresters, economic planners, and environmentalists. Stated simply, this chapter traces and represents the *mechanics* of this authority, and it does so in order, simultaneously, to subvert and replace (Bhabha, 1994: 22; see also Rose, 1995). It is written, then, not to *speak for* First Nations, but, rather, reflexively, in response to demands by Native peoples that, as non-Native Canadians, we learn to recognize how a colonial past continues to infuse a so-called 'post-colonial' present; how colonialist practices are not only part of an 'ugly chapter' in Canadian history, but are still *endemic* today, inscribed into the very ways that we visualize and apprehend the world around us.[5]

To explore these questions, forest industry literature, economic theory, and preservationist discourse are examined. In each, the persistence of colonialist practices in the present is pointed out, rather than focusing on how colonial relations were imposed in the past. What such an approach offers is an analysis of contemporary 'itineraries of silencing' which have contributed to the marginalization of Native voices in present-day resource development and land-use conflicts. More theoretically—but no less important—it also raises questions about how colonial power *operates* (i.e., in and through practices of representation) as opposed to how it is *wielded* (i.e., in juridical-political fields).

As Edward Said (1994: 7) has noted in his *Culture and Imperialism,* the 'struggle over geography' in nineteenth century imperialism was not only about 'soldiers and cannons', but also about 'ideas, about forms, about images and imaginings'; or, in other words, about representational practices that organized what was visible (and what remained invisible), and that, in turn, authorized the administration and regulation of foreign lands and life (see Mitchell, 1988). This remains the case today. Seen in the context of *contemporary* political struggles faced by British Columbia Natives, writing the 'war in the woods' as in part a *crisis of representation* does not reduce it to 'mere' philosophical or literary concerns. On the contrary, it insists on the political significance of representation. This is especially important as, today, Native groups must not only seek redress for past colonialist practices (dispossession of lands, the physical separation and segregation of Native peoples on reserves, the paternalistic administration of 'Native affairs' through the Indian Act, and so on), but must also confront a growing non-Native backlash that has girded itself in the seductive rhetoric of liberalism in order to question why Natives should be granted 'special privileges' to which everyone else is not 'entitled', or which might limit individual 'freedoms'.[6] The purpose of locating and representing the operation of neo-colonial rhetoric is precisely to disrupt them and to build other possible rhetorical and political spaces in which Native rights—based on cultural, political and spatial practices stretching from before contact, through the present, and extending into an open future—can be viewed as *legitimate,* rather than the complaints of spoiled children. In other words, Native peoples must today not only contest state and corporate jurisdiction over their traditional territories, they must also struggle against an array of contemporary representational strategies that legitimate non-Native rather than Native authority over these territories. As shown here, with each representation of the forest as *only* a 'resource landscape', each staging of the British Columbia economy as a sphere of production and exchange *unrelated to the colonial production of space*, each declaration of 'wilderness' as the absence of *modern* human presence, existing Native presence and Native rights are marginalized and, in a sleight of hand, the forest appears as an uncontested space of economic and political calculation: an entity without either history or culture, where no other claims than those of the 'nation' and its 'public' are seen to exist.

ABSTRACTING TIMBER, DISPLACING CULTURE: LEGITIMATING EXTRACTIVE CAPITAL IN THE TEMPERATE RAINFOREST

Modernity, since Marx, has been characterized by a maelstrom of change, whereby social life is continuously made subject to ever-new forces of 'creative destruction' and experienced as 'perpetual disintegration and renewal' (Berman, 1982: 15; Harvey, 1989). If this is true of *social* life, then its corollary must certainly be that *nature* in modernity has itself been subject to violent change, made and remade in the image of commodity production and technological change. Today this occurs on a global scale. From industrial agriculture and suburban sprawl to global warming and acid rain, the material landscapes of advanced capitalism have been reordered to such an extent that one writer has spoken of the 'end' of nature (McKibben, 1989) and, in one of the great ironies of the period, the preservation of nature necessarily follows the logic of the commodity form, such that 'ecological reserves' *produce* nature in the mirror image of capitalist production.[7] Canada's west coast differs from this general claim only in the predominance of a *single* commodity: timber. Perhaps captured best by the model of the 'normal forest' (see Figure 7.1— nothing more than the rationalization of *timber* production amid the heterogeneous, unruly, and culturally-infused temperate rainforest), the production of nature on the west coast of the continent centres so completely on the practices of scientific, industrial forestry—or, conversely, on the preservation of its mirror image, 'pristine nature'—that forest conflicts have tended to focus entirely on the *proportions* of the landscape dedicated to each.

In the face of this continual remaking of nature, ecologists have argued that in advanced capitalism, specific entities—constructed and given value as commodities-for-exchange within systems of signification and circuits of capital—are violently abstracted and displaced from their ecological surrounds, threatening the continued viability of ecosystems and wildlife populations. Indeed, this has become, perhaps, the most compelling 'green' critique of capitalist modernity—that nature becomes displaced into systems of signification, production, and exchange that have no 'intrinsic' relation to an underlying ecological order.[8] Without reducing the force of this critique (although its own rhetoric must not be allowed to escape scrutiny), the author wants to borrow its central insight, but turn it in a different direction. If industrial forestry on the coast has been guilty of abstracting the commodity from its *ecological* surrounds (with devastating consequences for local ecosystems, including hydrological cycles, fish stocks, and wildlife populations), then it has equally been involved in the abstraction of not simply 'timber', but the 'forest' (as a unit of production) from its *cultural* surrounds (with equally devastating effects on local Native communities). With this in mind, this discussion of contemporary 'itineraries of silencing' begins by turning to the representational practices of British Columbia's forestry industry and its strategies of legitimation.

CUSTODIANS OF THE FOREST

In many ways, MacMillan Bloedel (MB) has come to represent the corporate face of industrial forestry in British Columbia.[9] Perhaps for this reason it has been the subject of continual public scrutiny. Since the mid-1980s, MB—

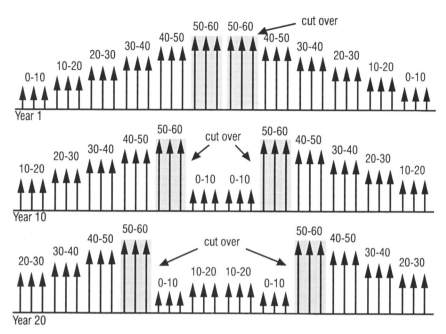

Remaking the forest in the image of the commodity. The 'normal forest' rationalizes forests in terms of the production of equal and sustained annual timber yields. Ecologists have argued that this abstracts the commodity from its ecological surroundings, threatening the continued viability of ecosystems and wildlife populations. Such models also take the forest to be only a physical entity, thereby abstracting trees from their cultural surroundings. In British Columbia this often results in the displacement of resource landscapes from existing Native territorialities. Reproduced with permission of David Demeritt.

Figure 7.1 Sustaining the yield in the normal forest

like most other forest companies operating in the province—has actively sought to legitimate its authority as the forest's 'custodian'. It has done so in a variety of ways—public relations pamphlets, television advertisements, visitor centres and forest tours being among the most visible. This section traces the mechanics of building corporate 'authority', and, in so doing, locates the simultaneous displacement and marginalization of Native voice that has been necessary for this authority to appear 'natural'.

MacMillan Bloedel builds its authority through carefully crafted representational strategies. Fully aware that legal arguments of tenure are inadequate (especially since most tenure-holdings in British Columbia are granted by the crown and not held in fee-simple), MB sets out to legitimate its authority by extending an invitation to evaluate its forest practices in terms of three criteria: *expertise, efficiency,* and *responsibility.* Expertise is conventionally demonstrated by appeals to science and technology. For instance, in the company's many glossy, attractively-packaged pamphlets, MB depicts a vast army of scientists, resource managers, and 'environmental specialists' shown working in the field, in the lab, or with computer simulations, while tables and graphs provide the reader with extensive 'technical' information about the forests and forest management (Figure 7.2).[10] Adapting new technology, readers are told, allows MB to "simulate the growth of present and future forests for 200 or more years and examine the results of different constraints" (MacMillan Bloedel, n.d. (2): 6).

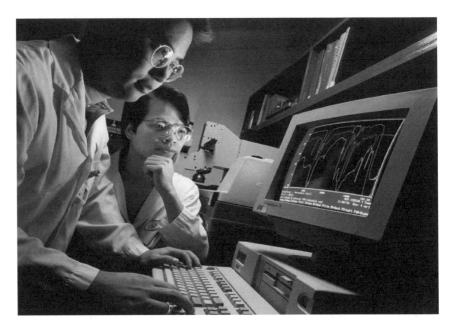

Figure 7.2   Technology and forestry

Likewise, the exhibits and interactive displays at the company's visitor centres in Port Alberni and Tofino describe MB's forest management cycle, invite visitors to 'test their knowledge' of temperate rainforest ecosystems, or, as in one computer game, interpellate visitors as corporate forest managers whose natural objective is to maximize profits through rational management practices (the *moral*, of course, is that this is possible only by understanding and accounting for 'nature'). In the case of MB's forest management cycle display, the focus is firmly on scientific expertise, documenting the careful assessment and monitoring necessary to develop ecologically sensitive 'site-specific' plans. The display informs visitors that after harvest "a site treatment plan is prescribed based on consideration of soil depth and texture, the amount of wood debris, and the species and health of small trees already growing on the site." Like a kind father guiding a child in the bewildering public spaces of the city, MB guides urban visitors through the alien spaces of the 'working forest'. Authority is, therefore, constructed through a comforting story of rational management, following a common practice in the 'modern' West of displacing questions of political legitimation from the realm of 'values' (or moral reason) to the realm of 'technique' (instrumental reason) whereby technical interests come to be established as 'values' in their own right (Habermas, 1972: 1987). MacMillan Bloedel, we are told, holds the most advanced knowledge and employs 'state-of-the-art' technologies; left to the company, the forest will be renewed, if not improved, for future generations.

MacMillan Bloedel also sets out to demonstrate that it is managing the resource in the public's best interest. This takes two forms: showing that the company obtains the greatest 'value' from the resource (efficiency); and, demonstrating that the company is responsive to non-timber forest values (responsibility). The first is accomplished by drawing direct links between MB's activities and the consumer demands and economic health of the province. MacMillan Bloedel explains that its operations are in a sense 'necessary' since it is simply meeting society's basic material needs by "grow[ing] and harvest[ing] trees and turn[ing] them into quality forest products that help satisfy society's need for communication, shelter and commerce" (MacMillan Bloedel, n.d.: 2). At other points, MB emphasizes its contributions to employment and government revenues, showing itself to be indispensable to the provincial economy. Likewise, in a recent series of television ads, the company responds to the criticism that it is more interested in achieving windfall profits (by liquidating forest resources and selling them as raw material or primary manufactured goods) than in contributing to the further development of the provincial economy. By focusing on new 'value-added' products (like Microllam and Parallam, laminated veneer lumber and parallel strand lumber respectively, which can use fibre from 'less-than-perfect-trees'), the company demonstrates that it is committed to obtaining ever new and increased 'value' from the forest resource, and, by extension, putting less pressure on 'old-growth' forests. MacMillan Bloedel, we are assured, is 'making the most of a renewable resource'. Finally, the company demonstrates its corporate responsibility. In its literature it notes that it facilitates and takes into account public input, opens 'its' forests to multiple uses, and far exceeds legislated responsibilities for the preservation of wildlife habitat. "The forests of British Columbia," readers are told, "are a great source of pride and concern for the people of the province. No one wants to see them decimated or devoted exclusively to timber production" (MacMillan Bloedel, n.d.: 12). Indeed, the corporation establishes itself as a 'disinterested' manager, mediating between the claims of various 'interest' groups: "The company's forestry policies are based on achieving an optimum balance for all users taking into account economic, recreational and environmental factors" (p. 12).

PUBLIC FICTIONS AND FICTIONAL PUBLICS: PRODUCING PURE SPACES OF ECONOMIC AND POLITICAL CALCULATION

Debates over forest management are vital, but equally significant is the manner by which MacMillan Bloedel's authority is built. By focusing exclusively on its forest management practices, MB carefully delineates the terms upon which its authority is to be evaluated. The open invitation to the 'public' to judge whether it has been a good *custodian* of the resource, for instance, displaces attention from its corporate interest (capitalizing forest resources) and from the province's tenure system (which provides multinational forestry companies with long-term access to the resource in the belief that lengthy tenure over large areas of forest land will be the most efficient, stable and sustainable organization of industrial forestry).[11] What interests me, however, is something quite different. Accepting MB's invitation to evaluate its *practices* as the basis for establishing authority is to fail to recognize the colonialist rhetoric that operates *through* these representational practices.

If MB's literature is read against the grain, it is possible to discern the subtle mechanics of its authority. MacMillan Bloedel's rhetoric of 'custodianship', for instance, can be seen to pivot on the mobilization of a potent (and often necessary) political fiction—the 'public'. Its authority appears legitimate because the company is seen to be meeting the standards of rational management, economic development, and ecological sustainability that the 'public' demands. As Bruce Robbins (1993) notes, in Western democracies the 'public' has often served as a rallying cry against private greed, propertied interests, and corporate and bureaucratic secrecy. But equally, it has served to silence 'minority' concerns. In this case, constructing and appealing to a 'public interest' serves MB well. Through it, the company is able to posit a *singular* body politic, situate the reader *within* it, and thus assume a *unified*, collective interest in the forest which all readers, on sober reflection, must share. This allows the company to draw an important equivalence: the health of the resource is by extension the health of the province and its 'citizens'. Legitimacy is thus solely an issue of who is the best manager of the resource.

This merits further attention. Not only does this rhetoric flatten out differences *within* British Columbia society—in this case rendering Native concerns either illegitimate, or, at best, only one of many 'special', and thus self-serving rather than 'common', interests—but it also enacts an important erasure that in many ways structures and enables the company's representational practices. What remains unmarked in MB's literatures, television advertisements, and visitor centres is a subtle, rhetorical manoeuvre: by displacing the 'forest' into the conceptual space of the 'public', MB abstracts the 'forest' and the 'land' from their specific cultural and political contexts and relocates them in the rhetorical spaces of the 'province' and the 'nation'. Such an abstraction is possible, of course, only if the forest can be made to appear as an unmarked, abstract category emptied of any social and cultural content. Indeed, in the Tree Farm License maps and satellite photographs found on the company's visitor centre walls and in its promotional literature, this is achieved: the forest appears as a purely 'natural' object without historical geographies. Ultimately, MB's forests are at once *any* forest and *no* forest at all: they appear simply as pure spaces of economic and political calculation. Emptied of *cultural* histories, the forest becomes a unit governed by *natural* history, and thus is free to be subsumed into a discourse of resource management (recently bound to a new, powerful meta-narrative of sustainability) and tied to the administrative spaces of the province, rather than to the local lifeworlds of its Native inhabitants. Indeed, it is this *absence*, rather than the positive discourse of scientific management, that makes MB's claim to 'custodianship' transparent. In this light, the 'normal forest' so dear to British Columbia foresters is much more than a model of the most 'rational' means of forestry; it is also perhaps the clearest articulation of the abstraction of 'timber' from its cultural surrounds and relocation within temporal and spatial logics that have no 'intrinsic' relation to local Native communities. As one critic has noted in relation to the Clayoquot Sound dispute, MB's forest planning maps impose an entirely different 'cultural geography' on the landscape, erasing and displacing already existing territorialities (Ingram, 1994).

THE ABSENT PRESENCE IN STAPLES THEORY

> *Race and ethnicity are frequently treated as 'add-ons' to political economy writings, as if these phenomena were peripheral to the way that Canadian society has unfolded historically and to the connections that have developed between the economic system and ideological, cultural, and political orders ... [Political economy accounts] recognize that a profound racism permeated the dynamics of both European-Native interaction and working-class formation in Canada, but they incorporate racism as a phenomenon of an exceptional 'ugly chapter' of Canadian history, rather than as a constituent and explanatory feature of Canadian historical development* (Abele and Stasiulis, 1989: 242, 244).

The second example is drawn from the supposedly dispassionate academy, and in particular, from economic *theory*. It is fashionable to dismiss the academy as an 'ivory tower' with tenuous links to the 'real world'. Yet, as Donna Haraway maintains, theory is anything but inconsequential. Indeed, as individuals we live social and economic theory daily, whether consciously or unconsciously. Amid the naturalization of social relations, *writing theory* is a necessary, albeit politically fraught practice which, Haraway suggests, produces "patterned visions" of "how to move and what to fear in the topography of an impossible but all-too-real present" (Haraway, 1992: 295). Haraway's statement can be read in two ways: theory is a *necessary* practice wedded to a political imaginary; and, theory *as* practice is 'situated' rather than universal. At once a tool by which to negotiate the present, it can never aspire to complete knowledge. At best, it provides a patterned vision, a lens.

For Canadian political economists, staples theory—developed by economic historian and communication theorist Harold Innis in the 1930s and 1940s— has provided a particularly valuable lens through which to understand the dynamics of Canadian economic and social life. Innis's writing and others inspired by it have been required readings for generations of Canadian students. Equally as important, staples theory has provided Canadian economic and social planners with maps by which to develop state industrial and economic strategy, as evidenced in the royal commissions and state policies of the 1970s and early 1980s (and the continued appeal to versions of staples theory by lobby groups like the Council of Concerned Canadians). Indeed, staples theory has been employed to address a wide array of contemporary political and economic questions: levels and forms of industrial development; regional disparity; degrees of corporate concentration; the amount and significance of foreign ownership; dependency (on foreign capital and on staples production); types and necessity of state intervention; levels of vulnerability to global economic change; and the production of specific resource and industrial landscapes.[12] However, to say that staples theory provides a 'patterned vision' is immediately to suggest that, like all theory, its explanatory value is rooted in specific concerns and in particular historical and institutional contexts. And, like all theory, its picture of the world gives the appearance of order, in part by drawing parameters around what is made 'visible' and what remains 'concealed'. Extending the visual metaphor, this section suggests that staples theory— especially as it has been applied in British Columbia—suffers from a colonial stigmatism that may require yet another corrective lens.

It is conventional to describe British Columbia's economy as a 'staples economy', based on the development and export of natural resources and primary manufacturing products like lumber, pulp and paper, and food goods (Shearer *et al.*, 1973; Bradbury, 1978; Hayter, 1978; Marchak, 1983; Davis and Hutton, 1989; Hayter and Barnes, 1990). Indeed, Roger Hayter and Trevor Barnes recently concluded that staples theory remains a useful 'heuristic' by which to understand not only British Columbia's economy, but the particular production of territory in the province (Hayter and Barnes, 1990). Their conclusions were based on a survey of British Columbia manufacturing, wholesaling, and producer service firms during the severe recession of the early 1980s. The authors found that, despite upheavals, British Columbia's 'global role' with respect to staples production was essentially unchanged from earlier periods. In the process of making their case, Hayter and Barnes articulate one of the clearest expressions of staples theory as a 'conceptual framework' by which to explain the British Columbia economy. In this section, Hayter's and Barnes's account is viewed through an 'anti-colonial' lens in order to locate how a colonial imaginary continues to structure the staging of British Columbia's space-economy. By interrogating their central categories, it becomes possible to locate an *absent presence* (the production of colonial space) that structures both economic relations in British Columbia, and the theories deployed to explain these. It is argued here that a colonialist imaginary is reproduced whenever economic development (both contemporary and historical) is seen as separate from, and unrelated to, this colonial production of space. However, this author's reading is in many respects sympathetic. What makes Hayter's and Barnes's model useful is that, while structured by a defining absence, its emphasis on institutions, technologies, and geography allows it to be reworked in ways that make it useful for a politics of decolonization.

THE INNIS TRIAD

Among the many reasons Hayter and Barnes find Innis's staples theory attractive is that it requires "an acute sensitivity to geographical and historical context" (p. 157). This is more than the disciplinary allegiances of two economic geographers. Rather, as many others have noted, staples theory has the advantage of taking seriously the specific conditions of Canadian economic development in ways that theories of development constructed with other regions in mind—namely the 'old world'—simply cannot. Innis remains attractive because he wrote theories *of*, and theories *from*, the margins of empire. Thus, over time, Hayter and Barnes argue, he developed a 'distinctively Canadian' theory of economic development. Innis's particular insight—by now 'common sense'—was that Canada had developed as a producer of staples products for metropolitan countries (initially Britain, later the United States), and that this had resulted, first, in particular institutional and geographical forms of Canadian economic development, and second, in a series of structural relations between global markets, institutions like the state and the firm, and technological development. Later writers would expand this into generalized theories of truncated industrial development and colonial dependency, the merits of which today remain highly debated.[13]

Leaving these aside, the focus of this chapter rests on what contemporary articulations of staples theory include and exclude, or, said differently, what this economic theory *of* and *from* the margins itself marginalizes. This focus relies on Hayter's and Barnes's recent attempt to describe staples theory schematically, in part because it is open to strategic adjustment. These writers suggest that Innis's staples thesis can be broken down into a *triad* of concerns: geography, institutions, and technology (Figure 7.3). Indeed, it is precisely this that makes Innis's theory sensitive to specific contexts. Within this triad, Hayter and Barnes argue, can be explained the formation of a distinct and unique Canadian *territory*, and, indeed, in Innis's own historical studies, all three elements are woven into complex stories in which no single one alone is

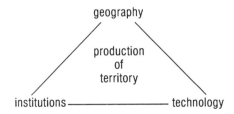

Figure 7.3   The Innis triad (after Hayter and Barnes 1990)

allowed to explain the specific historical and geographical forms of economic development.

Despite Innis's own careful work, critics like Abele and Stasiulis have found in Innis a strange omission. Innis's historical studies of the fur trade make much of Native agency, to the point where the trade emerges as a series of highly negotiated exchanges.[14] Yet, when he turns his attention to the present, Native people are absent. At one level this only seems reasonable; after the end of the fur trade, the settlement of the prairies, and the appearance of extractive capital, Native peoples are increasingly marginal to daily economic activity. In other words, they do not show up as actors—as labourers, as technological innovators, or as owners of the means of production. British Columbia historians and economists have generally followed the same path. Peter Ward (1980: 272), for instance, noted that "the Indians of British Columbia lost their central role in the economy of the region with the decline of the fur trade." Likewise, Patricia Marchak, in her study of the British Columbia forest industry, relegates the erasure of Native people to a short parenthetical statement: "Native peoples were pushed aside as their role in the history of furs became irrelevant" (Marchak, 1983: 32). At one level, both writers are correct—Native peoples were no longer 'visible' actors. But both writers accept the separation and segregation of Native people as a *fait accompli* after which the pressing issue (regarding Native peoples) becomes their *integration* into the workforce and associated problems of discrimination. What is not explicitly noted is that British Columbia's staples economy, rather than operating *without* and *apart* from Native participation, operates *on the basis of their absence*, and, further, that this absence was accomplished through the production of colonial space (about which more shortly). Following the production of a colonial spatiality in which indigenous populations were physically separated from 'resources', it no longer seems necessary to include Native peoples in order to theorize the British Columbia economy, which stands as a sphere of production and exchange apparently *unrelated* to Native peoples, who, at best, are now understood as an army of 'reserve' labour. Indeed, where Natives are 'written into' British Columbia (and Canadian) political economy, this generally follows three paths: historically as agents in the fur trade; as still present land-based producers (in peripheral regions); and as wage labourers. The corresponding political responses are, respectively, nostalgia, preservation, and struggles against employment discrimination. But to what extent does this adequately recognize the continuing role of Natives in the British Columbia economy, even if this role is, in essence, expressed in the negative, as an *absence*? What happens if we understand British Columbia's staples economies not solely as the product of technological change, physical geography, and the institutions of state and market, but as a social and historical product, where the form of the economy—and the possibility for it to operate in the way it does—is inseparable from the histories that have made these economic spaces and social relations of production possible?

Returning again to Hayter's and Barnes' model, if its triad of concerns are Innis's single most important contribution to Canadian economic theory, then it is also, as hopefully will be shown here, a triad that later writers have developed in ways that incorporate a series of buried colonial assumptions. In order to unpack this, it is necessary to pay attention to what is included in this 'triad', especially as applied to British Columbia. As outlined by Hayter and Barnes, all three elements are necessary. *Geography* is essential in two ways. First, *physical* geography is seen to make a difference in ways that orthodox economics—with its isotropic planes—rarely specifies. Innis's staples theory is insistently material, in part because the lay of the land, and the distribution of elements of the landscape, provides an initial template over which human activity is organized, but more than this, because at every turn the economic, the technological, and the geographical are wound together: territorializations are always co-productions among these constituent components. Second, Innis emphasized *spatial relations*, in particular, relations between the 'metropole' and the 'periphery'. These went some way to determining Canada's role in the global economy, and the particular forms of production found in the nation. The same can be said of British Columbia. Likewise, *institutions* could be divided into three distinct aspects: markets, firms, and the state. Innis insisted that all three were important, with the state—especially in staples producing regions like Canada —having an essential role for the regulation of land, the development of public works and infrastructure, and the financing of resource production. Finally, *technology* helped define the kinds of staples available in hinterland regions (indeed helped produce regions *as* hinterlands), and the types of staples demanded by the heartland. Not without reason, Innis has been seen as at times writing technological histories, where innovation appears to be the defining element; the production of territory on the margins of empire, after all, has always been tied closely to how technological innovation reworks spatial relations and gives rise to institutional responses. The introduction of specific technologies to the coastal forest industry, for instance, can be seen as central to the changing spatial patterns of forest operations, the centralization of industry and the dominance of a few large firms.

In the hands of economic geographers and historians, this broad model has been asked to bear the weight of explaining the development of British

Columbia's economy, something that Hayter and Barnes argue it does quite well. In many ways they are right; staples theory provides an acute sensitivity to the particularities of place—to explaining the 'good life' around Idaho Peak (Harris, 1985). Likewise, by developing this 'triad' of concerns, staples theory presents a theory of dynamic change grounded in the continually shifting constellation of technical, economic, institutional, and geographical relations, which provides a conceptual framework for the planner to deploy in projects directed towards social and economic regulation and the administration of 'populations'. As useful as this framework may be for charting the historic and future course of economic development, it provides little means by which to assess the colonial relations embedded in British Columbia economic development. By returning to Innis's 'triad', it is suggested that a very different 'patterned vision' can be articulated that allows other stories to be told about the development of British Columbia's staples economy and that reveals the *politics* of writing accounts of economic development.

ERASING THE PRODUCTION OF COLONIAL SPACE

If Hayter and Barnes, as geographers, are sensitive to the place of geography in staples theory, then they have a constrained notion of geography in mind. Geography, in this case, is limited to (a) physical geography and (b) spatial relations. No attention is paid to the *production of space* (except as an artifact *of* staples economies), and it is precisely this absence that allows staples theory to operate in a way similar to orthodox economic theory—as if space was only a matter of geometry, and not infused with either history or social relations. What is missing from this model (and from most recent accounts of British Columbia's staples economy) is the recognition that British Columbia's economy has been founded on the production of *colonial* space, by which is meant the division of the province's territory in the latter half of the nineteenth century into two distinct *orders* of space: one 'aboriginal' (primitive), delineated and contained within the 'reserve'; the other 'national' (modern), encompassing all that lay outside the bounds of the reserve (see Tenant, 1990 for a discussion of 'the Indian land question')—a division of space inscribed into the very administration of the nation and its lands. The administration of aboriginal space, for instance, is still vested in the federal Department of Indian Affairs, while the administration and regulation of 'crown lands' is located in specialized provincial ministries. The British Columbia Ministry of Forests, for example, has as its 'object' of administration provincial forest lands, which are seen as entirely separate (geographically and administratively) from Native communities. Indeed, this division of space and authority allows the naturalization of what might be called a colonial optic: a way of rendering each spatial domain visible *without reference to the other*. It is precisely such a colonizing vision that permitted JudgeSloan to propose a tenure system and the rationalization of the province's forests in a system of 'working circles' across virtually the entire extent of the province's territory. Without reference to local Native communities, the Sloan Commission in the 1940s inscribed a colonial spatiality at a scale never before encountered, and, in the process, allowed forestry— and its space-economies—to appear entirely separate from Native communities and existing territorialities. It is this representational logic that lay behind the increasing feeling of 'isolation' that Simon Lucas and the Nuu-chah-nulth experienced in the 1970s.

By 'staging' British Columbia's economy as solely a matter of physical geography and spatial relations, staples theory fails to account for how this organization of space continues to structure present-day economic development. Drawing on the earlier work of Shearer *et al.* (1973), Hayter and Barnes, for instance, outline the geographical, institutional and technological characteristics that gave rise to the 'long boom' which British Columbia's staples economy experienced after World War II. In terms of geography, they relate, resources were plentiful and cheap, foreign markets (USA) were expanding, and Vancouver was emerging as a city of metropolitan status; institutionally, state investment in infrastructure, long resource leases at low rates, and large externally-controlled corporations pursuing various integration strategies contributed to the province's economic growth and stability; and technologically, the province continued to rely on corporate research and development outside the province. What is peculiar about this potted history is not its focus on multiple geographical, institutional, or technological 'factors', but what *counts* as a factor. Certainly it makes sense to include the large unions as part of this institutional matrix, but if these, why not the juridical-political apparatus that until the 1950s criminalized obtaining funds to pursue land claims? This raises a fundamental question: what lies within the bound of 'the economy'? How are staples commodities (and economies) built? Is it solely a matter of physical geography and spatial relations? Why is it that resources were cheap and plentiful? Upon what basis did Vancouver emerge as a city of metropolitan status? For Hayter and Barnes, territory is seen to emerge from *within* Innis's matrix (which appears as the 'natural machinery' of the production of space); yet, is there not a production of space that *precedes* rather than follows from British Columbia's staples economy? Without asking this question, the 'economy' is allowed to stand autonomous from colonial legacies, as if operating *by* nature. The result—to rephrase Abele and Stasiulis—is that economic theories incorporate colonialism "as a phenomenon of exceptional 'ugly chapters' of Canadian history, rather than as a *constituent* and *explanatory* feature of Canadian historical development" (Abele and Stasiulis, 1989: 244, *italics mine*). Accordingly—evident not only in British Columbia but across the nation—efforts to deal with the 'underdevelopment' of Native communities have taken the form of 'appending' or 'incorporating' Native interests into an economy that, by its conceptual separation from its founding conditions, is presumed to pre-exist, rather than be built upon, Native marginalization.

Viewed through an anti-colonial optic, Innis's objective to 'write from the margins' appears ironic. Situated in Canada, looking back at the imperial centre, staples theory is resolutely anti-imperial, both in the tools it provides for understanding the margins of empire, and also in its resistance to the imperialism of metropolitan theory. Yet, when the gaze of staples theory is directed towards the territory of the Canadian nation, it is remarkably Eurocentric. Once Natives are no longer visible as active agents in the economy (after the end of the fur trade), they disappear. The economy operates as if *without* history.

With this in mind, Roy Vickers' painting *Vancouver* (Figure 7.4) can be read as an explicitly political articulation. The shadowy totem poles that loom behind the Vancouver skyline do not simply conjure nostalgic or romantic images of a long vanquished Native past. Rather they assert—in the very landscape that is built upon and erases colonial space—a continuing Native presence.

Figure 7.4   Vancouver: a political articulation

Indeed, placed alongside a recent widely-read paper by Davis and Hutton (1989) that outlined the changing dynamics and relations between British Columbia's 'two economies' (Vancouver and its hinterland), Vickers' images remind the viewer that the replacement of totem poles by skyscrapers was not a 'natural' development, but occurred, in part, through a silent colonial violence. In texts drawing on accounts of British Columbia as a staples economy, Vancouver is seen as emerging from its location between and relation to both a resource hinterland and global markets (Davis and Hutton, 1989; Hayter, 1978; Bradbury, 1987; Barnes *et al.*, 1992). This may be true, but what is never stated is that Vancouver's material landscape—including the skyline that Vickers' work places in question—incorporates as a condition of its possibility the isolation of Native people from the resource landscapes of the city's hinterland. In the face of white backlash to Native political expression, Vickers' image makes manifest the political unconscious of the province: the landscapes of the heartland are built upon an absent presence. Ironically, Innis said much the same of the metropole.

PRISTINE NATURE AND THE TROPES OF TRADITIONAL CULTURE

If Judge Sloan in the 1940s homogenized and rationalized a provincial 'forest' so as to make it available to industrial forestry, his vision of the forest and forestry has since been widely contested. Many critics have identified problems arising from the structure that the forest industry took on following Sloan's report, arguing essentially that too much control has been centralized in too few actors (see Drushka *et al.*, 1993).[15] Indeed, in this vein, debate has turned on the relations between tenure (private vs. public; corporate vs. community), corporate concentration, forest practices, levels of employment and, ultimately, who *benefits* from the present structure of the industry (local communities, state ministries, corporations, labour). But the most vociferous critics—and arguably the most successful in terms of public opinion—have been the province's environmental groups, which have consistently resisted industrial forestry for its transformation of physical landscapes. Against the remaking of nature solely on the basis of economic rationality, they have asserted an *ecological* rationality that visualizes 'nature' as consisting of a series of interrelated 'systems', where the *disruption* of any one results in reactions and modifications throughout the 'web' of relations that link local ecosystems with a global biosphere. Indeed, since Rachel Carson's *Silent Spring* (1962), ecosystem theory has provided a powerful and remarkably flexible set of concepts that have radically reconfigured the production and politics of place and nature in Western cultures.

In British Columbia, as in other places where extractive capital has extended its reach into traditional territories of aboriginal peoples, environmentalists and Native groups have forged important alliances. However, such coalition building is, almost by definition, fragile. As indicated in this final example, Native peoples on Canada's West Coast have good reason to be wary of the representational practices of the environmental movement, for whom 'nature' is often understood in radically *different* ways than by Native groups. Certainly, many perceived benefits have resulted from cooperation with environmental groups (especially in forestalling logging and asserting Native land rights; notably, Meares Island, South Moresby, Stein Valley), and many individuals and groups in the environmental movement are strongly committed to an anti-colonial politics. Yet, it is not immediately clear that the environmental movement's concerns for *preservation* can be mapped onto Native land claims or, for that matter, the social and economic ambitions linked to these claims. At certain times, the rhetoric of the environmental movement is resolutely neo-colonial.

WILDING NATURE

NorthAmericanenvironmentalism brings together and often combines two visions of nature: wilderness (nature as absence of culture) and ecology (nature as system). In their application in British Columbia, both present problems.

As Ramachandra Guha (1989) and William Cronon (1995) have noted, radical American environmentalism is fundamentally preoccupied with wilderness, in many ways a legacy of European settlement of the 'frontier'. Indeed, the desire to preserve 'wilderness' emerged just as the 'frontier' was understood to be 'closed' (Nash, 1967). For many in the environmental movement, wilderness is at once ontological (an objective thing to be *saved*) and ethical (a moral call to *let be*). Yet, as Patricia Jasen (1995) has recently shown in her brilliant discussion of nature tourism in Ontario, what counts as wilderness is not so much an essence as a *relation*. It is heavily invested by, and tied to, structures of *desire* for Western urbanized subjects (see also Cronon, 1995). Indeed, as she notes, this is achieved through representational practices that divide the world in two between the modern/industrial/humanized and the pre-modern/primitive/pristine. Wilderness thus becomes that place where the former has not fully infected the latter, and the latter—scripted through Edenic narratives of a natural harmony before the 'fall'—becomes freighted with moral values and cultural symbols.

Wilderness is thus less a physical place than a semiotic relation: it is those sites that can most *appear* as characterized by the *absence of culture*, or, in other words, those areas which show few signs of human 'modification' and thus promise unmediated experience with *primal* nature, that can be coded as 'wild'. In short, wilderness represents a flight from history—'wild' places exist outside time. Represented thus, what appears as industrial society's 'other' becomes a terrain across which a whole series of metropolitan anxieties can be played out: including, as Jasen demonstrates, concerns over stress (brain-fag) and physical deterioration in late nineteenth-century urban cultures, and today, as is evident in the growing popularity of adventure travel, anxieties over the technologization and commodification of 'life', a most perplexing condition which seems to threaten what it means to be truly *human* (and, often, what it means to be *male*). What makes wilderness an object of desire, therefore, is a relation of *difference*: the further a particular site can be made to appear from the unauthentic, denaturalized spaces of the city, the easier to appropriate it into a desire to rediscover and recover the lost essence of the 'human' and a world before the 'fall'. Wilderness is the 'pure' and the 'uncontaminated'.[16] And, as middle-class white Americans have learned, in wilderness—from Henry David Thoreau to Edward Abbey—lies the path to wholeness.

Ecosystem ecology has provided other metaphors that are equally implicated in positing nature as outside history. At one level, the science of ecology is little more (but no less) than the mapping of flows of energy and matter, and thus makes no differentiation between 'culture' and 'nature'. Yet, this is hardly how the environmental movement has deployed the language of ecosystem ecology. Instead, in North American environmental literatures, ecosystems are almost invariably represented as stable, self-regulating 'natural' systems, external to human societies. As Demeritt (1994) has noted, the romantic metaphors that permeate ecosystem ecology—balance, equilibrium, harmony—seem ready-made to highlight human disturbance.[17] Thus, these 'natural' systems appear everywhere to be *under threat* by humans. This introduces a peculiar and highly problematic logic: visible signs of human presence imply the modification— and thus destruction—of a pre-existing 'pristine' system, and, following the logic of cybernetics, imply a series of further ramifications on both local and global levels.

It is important to note that there is no *necessary* relation between romantic notions of 'wilderness' and the rhetoric and principles drawn from the science of ecology. As both David Harvey (1994) and Andrew Ross (1994) have noted, New York City is an ecosystem just as is a putatively uninhabited atoll in the South Pacific. Yet, by removing humans from ecosystems, notions of 'wilderness' and 'ecosystem integrity' are drawn into close proximity, such that in both rhetorics, what is seen as the most fundamental problem is nature's humanization. In both nature is 'wilded'—it is that which pre-exists culture.

THE ABSENCE OF THE 'MODERN':
NEO-COLONIAL TROPES IN THE RAINFOREST

Rhetorics of wilderness are problematic in two respects. First, as Cronon (1995: 80) persuasively argues, its flight from history represents the false hope of an escape from responsibility. It provides "the illusion that we can somehow wipe clean the slate of our past and return to the tabula rasa that supposedly existed before we began to leave our marks on the world." Our very presence in nature represents its fall, thereby displacing attention from urgent questions over what *kinds* of marks we wish to leave on this world to nostalgic desires for the preservation of the 'primal' and 'pristine'.[18] Second, the escape from history that is at the core of wilderness is deeply problematic for Natives, since it is their traditional territories that often become caught in the semiotic webs of 'wilderness'. The trouble with wilderness becomes evident immediately in the representational practices of west coast environmental groups in whose literature nature is framed such that only certain things are allowed to appear within the visual field. This has been particularly evident in various publications of 'nature' photography where the lens of the camera has been trained in such a way as to construct scenes of west coast natures that appear 'pristine'. At one level, this provides the environmental movement with its emotional appeal: nature is *either* 'pristine' or 'spoiled', there is no middle ground (the sexualization of this imagery is immediately apparent, more so in the not-too-infrequent equation of the body of nature with the eroticized female body). The message is clear: like masculinist constructions of virginity, once spoiled never again pure. Yet, as is now well known, west coast lands are *not* unoccupied wilderness and haven't been for millennia.

As discussed elsewhere in more detail (Willems-Braun, forthcoming), west coast nature photography is marked by a tension between representing 'wilderness' and representing a Native presence. Adrian Dorst's spectacular(ized) photographs of Clayoquot Sound are instructive in this respect. Despite depicting an area inhabited (and extensively modified) by a large Native population, Dorst's photos rarely admit the presence of Native peoples. More importantly, when Native presence appears within the frame, it does so *only* under the guise of the 'traditional' (see Figure 7.5). In many ways, this is unfortunate.

Figure 7.5   A "traditional" Native presence

Dorst's photographs present compelling reasons (aesthetic and ecological) for resisting the intrusion of industrial forestry, yet by containing Native presence within the rhetorics of the 'traditional', they simply update deeply held, highly romantic, and resolutely Eurocentric notions of the 'ecological Indian', where 'traditional' practices are assumed to be 'ecologically harmonious'. *Modern* Native activities, and the use by Native peoples of what are taken to be modern (European) technologies, never appear in the pages of these books, nor, for that matter, elsewhere in the literature of British Columbia's environmental movement. Technology is equated with an intrusive modernity which imperils wilderness, and that, in the hands of 'traditional' Natives appears 'inauthentic'. This is a disempowering logic that belies the support for Native land rights which has been articulated by environmental groups such as the Western Canada

Wilderness Committee and Greenpeace. Such rhetoric—whether intended or not—equates 'traditional' Natives *with* nature, narrowly delineating the positions that Native peoples are able to occupy within forestry debates, and presenting Native peoples with an imperative: resist modernization or risk losing both your identity and your voice! By equating 'wilderness' with the absence of modern culture, and representing First Nations only through the lens of the 'traditional', the British Columbia environmental movement risks a subtle imperialism that denies Native peoples voice as *modern cultures* that are not solely interested in preserving what, at the end of the day, is little more than a 'mirror-image' of industrial production and the object of a middle-class urban desire: wilderness. Native culture becomes encased in the same museum exhibit as nature.[19] For British Columbia First Nations, situated in the hinterlands of a global staple trade, industrial forestry in itself is not the problem; rather, it is who has authority over the *production of nature* in their traditional territories.

POSTSCRIPT: ARTICULATING SOCIAL NATURE

In this chapter, an attempt is made to link the 'isolation' that Simon Lucas expressed to the Pearse Commission in 1975, and that is still widely experienced by many Native communities, with the 'itineraries of silencing' that have marginalized Native voices in debates over rights of access and responsibilities of use in British Columbia's rainforests. What has been shown, in the three examples chosen, is that Native peoples must negotiate a representational terrain marked by colonialist practices. Of course, not all foresters, staples theorists, and environmentalists operate under the thrall of these representational logics, yet they remain prevalent ways of framing the 'forest', its 'custodians', its 'publics' and its 'economies'. To leave the story here, however, would be to represent only the effectiveness of colonial power, without recognizing practices of resistance. As evident in the many court challenges and political actions of Native peoples, such representational logics have been challenged on numerous fronts. This discussion concludes by providing more detail to an example of what might be called a contestatory politics of *articulation* which—like the Vickers painting discussed earlier—disrupts the 'itineraries of silencing' found in colonialist rhetorics and articulates alternative rhetorical spaces.

Over the past number of years, the work of Coast Salish artist Yuxweluptun (Lawrence Paul) has been widely exhibited, most recently in a one-person show at the Morris and Helen Belkin Art Gallery at the University of British Columbia. Trained at the Emily Carr School of Art, Yuxweluptun has forged the sort of hybrid vision that is so unsettling to Western histories of art and artifacts. Combining the cosmologies and visual motifs of North West Coast culture with the modernist tradition of surrealism, Yuxweluptun's work is sufficiently aesthetically sophisticated by 'Western' standards (i.e., not 'primitive') to garner considerable critical attention, yet resolutely grounded in the tradi-

tions and politics of west coast Native life. Already an industry of critique has
begun to weave stories around his art, debating its relation and points of inter-
section with traditions of 'West Coast Indian Art' and 'European modernist'
traditions, questioning its position vis-à-vis the institutions of artistic production in Canada, and worrying over the wedding of aesthetics and politics.[20]

For the purposes of the present discussion, however, what is intriguing
about Yuxweluptun's 'salvation art' (the phrase is his), is the manner in which
his art unselfconsciously combines politics and aesthetics in a landscape vision
that destabilizes accepted renderings of nature on the west coast. As Charlotte
Townsend-Gault (1995) notes, Yuxweluptun paints not land*scapes* but land
claims. If Emily Carr—still mythologized as the 'artist of the West Coast'—
purified coastal landscapes into two elements: the decaying artifacts of a dy-
ing Native culture and a brooding, spiritualized, all-encompassing nature,
Yuxweluptun does the opposite. Nature in his work does not overwhelm a
vanquished culture; rather, the 'natural' landscapes that he paints are *built
from* west coast Native motifs. His *Scorched Earth, Clear-Cut Logging on Native
Sovereign Lands, Shaman Coming to Fix* (see Figure 7.6), for instance, is unapolo-
getically political on a number of levels. On the one hand, it can be read as a
critique of the ecological violence of industrial forestry. The land weeps, as do
the sun and a human figure resting on a distant hill. But this painting articulates much more, for as Townsend-Gault notes, Yuxweluptun does not simply paint a 'green' critique, but asserts Native *rights* to a land being destroyed.

Figure 7.6   Yuxweluptun's 'salvation art': *Scorched Earth, Clear-Cut
                  Logging on Native Sovereign Lands, Shaman Coming to Fix*

The various components of the landscape itself—trees, mountains, the sun, even tree-stumps—are constructed out of Native motifs. There is no nature *apart* from these. Further, it is from within native culture—rather than yet another intervention by non-native authorities—that this violence is healed. Similarly, *Clayoquot* (Figure 7.7) depicts the violence of clearcuts as the interruption of Native culture, as gaping wounds in a cultural landscape, its square rational spaces gouged into hills formed from Coast Salish ovoids.

Figure 7.7   Yuxweluptun's 'salvation art': *Clayoquot*

Yuxweluptun articulates a landscape politics that resists incorporation within a colonialist visuality, an image of the land that refuses to let it settle as only a resource landscape, as only wilderness, or as an entity without a constitutive Native presence. His vision is of a social nature, one without pure essences: one where native presence and native authority must be seen to exist. His landscapes are dynamic and fully historical—sites of cultural and political contestation, where Native identity and the physical environment are never 'fixed' in a timeless relation, but part of a complex dialectic of society and nature. Far from being *outside* history—or simply empty spaces available to economic and political calculation—these landscapes are at once fully Native and fully modern, but still deeply ecological.

ENDNOTES

[1] Pearse Commission public hearings, Victoria, October 30, 1975. Until the 1980s, anthropologists referred to the indigenous peoples of the west side of Vancouver Island as the "Nootka", an appellation commonly attributed to Captain Cook. The West Coast District Council of Indians Chiefs is now known as the Nuu-chah-nulth Tribal Council, and consists of representatives of the council's 14 member tribes from the west coast of Vancouver Island.

[2] In a report released six months after the decision , the provincial ombudsman claimed that the Nuu-chah-nulth had not been adequately consulted in the events leading to the development and release of the controversial land-use plan (Office of the Ombudsman, 1993). Following the release of the report, an interim agreement was signed between the central region tribes of the Nuu-chah-nulth Tribal Council and the provincial government that incorporated the Nuu-chah-nulth as co-managers of certain regions. Interim agreements are increasingly prevalent as First Nations and the provincial government seek ways to manage resources on the traditional territories of various First Nations while treaty negotiations are ongoing.

[3] It *is* true, as Nettie Wild (1993) has shown in her controversial film *Blockade*, that Native peoples—as individuals and communities—have actively participated in the transformation of their traditional territories by industrial forestry. How this is documented, however, matters greatly. If allowed to stand alone—without noting the complex dynamics that inform the lives of Natives in British Columbia (from physical segregation to contemporary economic dependency), and without sufficient attention paid to the romantic assumptions of 'ecological' Indians that lurk beneath the surface—such statements risk providing a basis on which to dismiss the legitimate concerns of contemporary First Nations.

[4] Indeed, since contact First Nations on the West Coast have routinely contested the imperial and colonial rhetorics that abstracted and displaced resources from their local social and cultural relations (see Clayton [in press] for an account of these displacements). In recent years, this has taken the form of blockades, court challenges, film making, intergovernmental negotiations, contestatory practices in West Coast Native art, and political activities on a multitude of other fronts.

[5] An excellent discussion of how North American culture remains infused with colonialist practices is found in Shohat and Stam, 1995.

6   This rhetoric has become prevalent in policy statements issued by the British Columbia Liberal Party and the federal Reform Party (i.e., "one people, one law").

7   In *The Mirror of Production*, Baudrillard (1975) shows how Marxist theory cannot escape the orbit of a bourgeois capitalist imaginary. In a similar way, green movements are irrevocably (and necessarily) tied to systems of nature's production in industrial capitalism. Thus, the preservation of nature comes to mirror the relentless commodification of nature as capital stalks the earth in search of surplus value. Indeed, one of the ironies of the environmental movement's attempts to preserve 'wilderness' is that it can do so only through the language of 'value' (see Demeritt, 1996).

8   It can be argued that all cultures, and not only those of advanced capitalism, displace nature into abstract systems of commodities and exchange (see Appadurai, 1986, for an excellent discussion of commodities and culture).

9   That MacMillan Bloedel has come to represent corporate forestry in British Columbia is the result of a number of factors: it is one of the only large corporations working in British Columbia's forests today that began as a local firm (and thus is highly identified with the province), and closely tied to various stages in the development of the industry, including its internationalization); its forest tenures include some of the most spectacular stands of 'old growth'; these same tenures are accessible to Vancouver residents; and, it has at different times publicly challenged environmentalists, resulting in considerable media exposure.

10   I explore MacMillan Bloedel's representational strategies in greater detail in Willems-Braun (forthcoming).

11   Public sustained yield units (PSYUs) are leased for 21 years, and can be renewed subject to the lessee meeting performance standards. There have been very few instances where renewal applications have been rejected.

12   For some of the many highly contested uses of staples theory in Canadian political economy, see the volume edited by Clement and Williams (1989).

13   Important texts in these debates include: Levitt, 1970; Naylor, 1975; Clement, 1977; Britton and Gilmour, 1978; Laxer, 1985; Williams, 1986; Bradford and Williams, 1989.

14   This approach is developed much further by Ray, 1974, 1978; Trigger, 1985; and Thistle, 1986. For British Columbia see Fisher, 1977. For a more detailed discussion of a period characterized by a highly negotiated 'middle ground' between Natives and Europeans in the Great Lakes region, see White, 1991.

15   Indeed, the Pearse Commission had already in 1975 come to many of the same conclusions. However, forest tenure has remained a most contentious issue, one which no government has been willing to address in a comprehensive fashion.

16   As has been frequently noted, the same nostalgic rhetoric has been inscribed not only on the body of 'nature', but on the 'native' body, which, in its signs of difference from we moderns, promises the same recovery of not-quite-yet-lost communion with primal (human) nature. This is also why the 'modern' native, dressed in 'Western' clothing or, worse, consuming 'Western' commodities, is so often a disappointment to Western travellers (see Zurick, 1995).

17   The rise of 'dynamic ecology' in the 1980s which emphasizes disequilibrium and disturbance has placed in question the 'scientific' basis of much romantic environmentalism, yet appears to have resulted in little change in the rhetoric used by environmental groups, nor the way that science of ecology is enrolled in projects of 'saving' nature. See Botkin (1990) and Worster (1990) for accounts of the 'ecology of chaos'.

18   From such rhetoric flows programs of action, and one only needs to compare the vast resources directed by British Columbia environmental groups towards the defence of

wilderness with the meagre resources directed towards producing healthy urban environments (or towards ecologizing forestry) to discover how 'wilderness' operates ideologically, and, perhaps more perniciously, becomes implicated in class relations. "Wilderness," Cronon (1995: 81) writes, in a statement certain to raise the ire of preservationists, "poses a serious threat to responsible environmentalism at the end of the twentieth century."

[19]   This mirrored in 'official' environmental literatures and not only nature photography allied with the movement. Environmentalists frequently refer to culturally modified trees and to the importance of the forest to 'traditional' Native cultural and spiritual practices. Thus, when the Sierra Club interprets Nuu-chah-nulth resistance as a call 'for the permanent protection of their cultural and spiritual values, environmental subsistence sites, marine resources, and salmon bearing streams and rivers,' they foreclose on an 'open future' for the Nuu-chah-nulth on their traditional territories (Sierra Club, n.d.).

[20]   See the catalogue that accompanied the Belkin exhibit (Yuxweluptun, 1995).

REFERENCES

Abele, F., and Stasiulis, D. (1989). Canada as a 'White Settler Colony': What about natives and immigrants. In W. Clement and G. Williams (eds.), *The new Canadian political economy* (pp. 240-77). Montreal & Kingston: McGill-Queens University Press.

Appadurai, A. (ed.) (1986). *The social life of things: Commodities in cultural perspective.* Cambridge: Cambridge University Press.

Barnes, T., Edgington, D., Denike, K., and McGee, T. (1992). Vancouver, the Province and the Pacific Rim. In G. Wynn and T. Oke (eds.), *Vancouver and its region* (pp. 200-33). Vancouver: UBC Press.

Baudrillard, J. (1975). *The mirror of production,* translated by M. Poster. St. Louis: Telos Press.

_____ (1984). Symbolic exchange and death, translated by Charles Levin. In J. Fekete (ed.), *The structural allegory. Theory and history of literature, Vol. II.* Minneapolis: University of Minnesota Press.

Berman, M. (1982). *All that is solid melts into air: The experience of modernity.* London: Verso.

Bhabha, H. (1994). *The location of culture.* London: Routledge.

Botkin, D. (1990). *Discordant harmonies. A new ecology for the twenty-first century.* New York: Oxford University Press.

Bouchard, D., and Vickers, R.H. (1990). *The elders are watching.* Tofino: Eagle Chance Enterprises Ltd.

Bradbury, J. (1978). The instant towns of British Columbia: A settlement response to the metropolitan call on the productive base. In L.J. Evenden (ed.), *Vancouver: Western metropolis* (pp. 117-33). Victoria: University of Victoria.

_____ (1987). British Columbia: Metropolis and hinterland in microcosm. In L.D. McCann (ed.), *Heartland and hinterland: A geography of Canada* (pp. 400-440). Scarborough: Prentice Hall.

Bradford, N., and Williams, G. (1989). What went wrong? Explaining Canadian industrialization. In W. Clement and G. Williams (eds.), *The new Canadian political economy* (pp. 54-76). Montreal & Kingston: McGill-Queens University Press.

Britton, J., and Gilmour, J. (1978). *The weakest link: A technological perspective on Canadian industrial underdevelopment.* Ottawa: Science Council of Canada.

Carson, R. (1962). *Silent spring.* Boston: Houghton Mifflin.

Clayton, D. (1995). *Island of truth: Vancouver Island from Captain Cook to the beginning of the colonial period*. Ph.D. Dissertation. Department of Geography, University of British Columbia, Vancouver.

Clement, W. (1977). *Continental corporate power: Economic elite linkages between Canada and the United States*. Toronto: McClelland and Stewart.

Clement, W., and Williams, G. (eds.) (1989). *The new Canadian political economy*. Montreal & Kingston: McGill-Queens University Press.

Cronon, W. (1995). The trouble with wilderness; or getting back to the wrong nature. In W. Cronon (ed.), *Uncommon ground: Toward reinventing nature* (pp. 69-90). New York: W.W. Norton.

Davis, H.C., and Hutton, T. (1989). The two economies of British Columbia. *BC Studies, 82,* 3-15.

Demeritt, D. (1994). Ecology, objectivity and critique in writings on nature and human societies. *Journal of Historical Geography, 20,* 22-37.

_____ (1996). *Knowledge, nature and representation: Clearing for conservation in the Maine woods*. Ph.D. Dissertation. Department of Geography, University of British Columbia, Vancouver.

Derickson, H. (1991). *Native forestry in British Columbia: A new approach. Final report*. Victoria: Task Force on Native Forestry.

Drushka, K., Nixon, B., and Travers, R. (eds.) (1993). *Touch wood: B.C. forests at a crossroads*. Madeira Park: Harbour Publishing.

Fisher, R. (1977). *Contact and conflict: Indian-European relations in British Columbia, 1774-1890*. Vancouver: UBC Press.

Guha, R. (1989). Radical American environmentalism and wilderness preservation: A third world critique. *Environmental Ethics,* 11(1): 71-84.

Habermas, J. (1972). *Knowledge and human interests*. London: Heinemann.

_____ (1984). *The theory of communicative action, vol. 1,* translated by James McCarthy. Boston: Beacon Press.

_____ (1984). *The theory of communicative action, vol. 2,* translated by James McCarthy. Boston: Beacon Press.

Hall, S. (1996). When was the 'post-colonial'? Thinking at the limit. In I. Chambers and L. Curti (eds.), *The post-colonial question: Common skin, divided horizons* (pp. 242-60). London: Routledge.

Haraway, D. (1992). The promises of monsters: A regenerative politics of inappropriate/d others. In L. Grossberg, C. Nelson, and P. Treichler (eds.), *Cultural studies*. London: Routledge.

Harris, C. (1985). Industry and the good life around Idaho Peak. *Canadian Historical Review,* 66(3), 315-43.

Harvey, D. (1989). *The condition of postmodernity: An enquiry into the origins of cultural change*. Oxford: Blackwell.

_____ (1993). The nature of environment: The dialectics of social and environmental change. *The Socialist Register,* 1993, 1-51.

Hayter, R. (1978). Forestry in British Columbia: A resource basis of Vancouver's dominance. In L.J. Evenden (ed.), *Vancouver: Western metropolis* (pp. 95-115). Victoria: University of Victoria.

Hayter, R., and Barnes, T. (1990). Innis' staple theory, export and recession: British Columbia, 1981-1986. *Economic geography,* 66(2), 156-73.

Ingram, G.B. (1994). The ecology of a conflict. In R. Hatch (ed.), *Clayoquot and dissent*. Vancouver: Ronsdale.

Jasen, P. (1995). *Wild things: Nature, culture and tourism in Ontario, 1790-1914*. Toronto: University of Toronto Press.

Latour, B. (1993). *We have never been modern*, translated by Catherine Porter. Cambridge: Harvard University Press.

Laxer, G. (1985). The political economy of aborted development. In R. Brym (ed.), *The structure of the Canadian capitalist class*. Toronto: Garamond Press.

Ledyard, J. (1783). *A journal of Captain Cook's last voyage to the Pacific Ocean, and inquest of a North-West Passage between Asia and America*. Hartford.

Levitt, K. (1970). *Silent surrender: The multinational corporation in Canada*. Toronto: Macmillan.

McKibbon, B. (1989). *The end of nature*. New York: Random House.

MacMillan Bloedel (n.d.). *Beyond the cut: MacMillan Bloedel's forest management program*. Vancouver.

_____ (n.d.) [2]. *Future forests*. Vancouver.

Marchak, P. (1983). *Green gold: The forest industry in British Columbia*. Vancouver: UBC Press.

Mitchell, T. (1988). *Colonizing Egypt*. Berkeley: University of California Press.

Nash, R. (1967). *Wilderness and the American mind*. New Haven: Yale University Press.

Naylor, R.T. (1975). *The history of Canadian business 1867-1914*. 2 vols. Toronto: James Lorimer.

Office of the Ombudsman of British Columbia (1993). *Administrative fairness of the process leading to the Clayoquot Sound decision*. Report No. 31. Victoria.

Pearse, P. (1975). *Timber rights and forest policy in British Columbia*. 2 vols. Royal Commission on Forest Resources. Victoria: Queen's Printer.

Ray, A. (1974). *Indians in the fur trade: Their role as hunters, trappers and middlemen in the lands southwest of Hudson Bay 1660-1870*. Toronto: University of Toronto Press.

Ray, A., and Freeman, D. (1978). *'Give Us Good Measure': An economic analysis of the relations between the Indians and the Hudson's Bay Company before 1763*. Toronto: University of Toronto Press.

Robbins, B. (1993). Introduction: The public as phantom. In B. Robbins (ed.), *The phantom public sphere*. Minneapolis: University of Minnesota Press.

Rose, G. (1995). The location of culture. *Environment and Planning D: Society and Space*, 13(3), 365-76.

Ross, A. (1994). *The Chicago gangster theory of life: Nature's debt to society*. London: Verso.

Said, E. (1994). *Culture and imperialism*. New York: Vintage Books.

Shearer, R., Young, J., and Munro, G. (1973). *Trade liberalization and a regional economy: Studies of the impact of free trade on British Columbia*. Toronto: University of Toronto Press.

Shohat, E., and Stam, R. (1994). *Unthinking Eurocentrism: Multiculturalism and the media*. London: Routledge.

Sierra Club (n.d.). *Clayoquot Sound: Wilderness splendour*. Victoria: Sierra Club of Western Canada.

Tenant, P. (1990). *Aboriginal peoples and politics: The Indian land question in British Columbia, 1849-1989*. Vancouver: UBC Press.

Thistle, P. (1986). *Indian-European trade relations in the Lower Saskatchewan River-region to 1840*. Winnipeg: University of Manitoba Press.

Townsend-Gault, C. (1995). The Salvation Art of Yuxweluptun. In *Lawrence Paul Yuxweluptun: Born to live and die on your colonialist reservations*. Vancouver: UBC Press.

Trigger, B. (1985). *Natives and newcomers*. Montreal & Kingston: McGill-Queens University Press.

Ward, W.P. (1990). Class and race in the social structure of British Columbia. *BC Studies*, 45 (Spring'76), 17-35.

White, R. (1991). *The middle ground: Indians, empires and republics in the Great Lakes Region, 1650-1815*. Cambridge: Cambridge University Press.

Wild, N. (1993). *Blockade*. Montreal: National Film Board of Canada [film].

Willems-Braun, B. (forthcoming). Buried epistemologies: The politics of nature in 'post' colonial British Columbia. *Annals of the Association of American Geographers*.

Williams, G. (1986). *Not for export: Toward a political economy of Canada's arrested development*. Toronto: McClelland and Stewart.

Worster, D. (1990). The ecology of order and chaos. *Environmental History Review*, 14, 1-18.

Yuxweluptun, L.P. (1995). *Lawrence Paul Yuxweluptun: Born to live and die on your colonialist reservations*. Vancouver: UBC Press.

Zurick, D. (1995). *Errant journeys: Adventure travel in a modern age*. Austin: University of Texas Press.

SECTION II
Industry in Transition

Globalization of Canada's Resource Sector: An Innisian Perspective

Bruce W. Wilkinson

Department of Economics, University of Alberta

INTRODUCTION

Globalization and rapid global change are the order of the day. This chapter will initially highlight a number of the key characteristics of these global forces and consider their relation to and implications for the forestry sector, the main focus of this volume. Because this volume is one of a series of books being published in honour of Harold Innis, an Innisian perspective will be taken when examining the global developments and their ramifications. In the process, a number of the insights Innis had, which are relevant to Canada and to forestry, will be introduced.

GLOBALIZATION: ITS NATURE

The essence of globalization is that corporations of all sizes, even the small ones, see themselves as actors on the world stage, not merely as stagehands at the local little theatre group, and nations in turn see themselves as integral parts of the evolving world economy, not islands unto themselves. As a consequence, production, technology, consumer tastes, and standards all are becoming internationalized, with liberalized international, intra-industry trade and capital flows linking the markets of the world. Governments, facing pressures to privatize and deregulate and liberalize, increasingly find themselves with diminished control over the processes and strategies of production, except insofar as they can represent the vested interests of their corporations in the development of international rules of corporate behaviour and/or privilege. Constant change and adaptation become the norm, and knowledge of diverse types becomes the crucial ingredient of continuing competitiveness. In this scenario, services of a wide variety of types—educational, research, consulting, communication, transportation, and financial—become of profound importance to the goods-producing sectors and as sources of employment; communication technology is expanding exponentially; the educated person becomes one who knows how to continue to learn rather than someone who achieved some degree or certificate at a point in time; and the unskilled, limited-education blue collar worker as well as many others in the middle class, often face adverse income and employment situations for which they are ill-prepared (Courchene, 1994; Drucker, 1994; Harris, 1993; and Hart, 1993).

From these characteristics a tendency exists to take for granted that the only industries which will be able to survive and prosper are the so-called high-tech manufacturing and service industries, and only those nations which are successful in promoting the advancement of such industries by whatever means are still available to them, will flourish.[1] Accordingly, the resource-producing industries and the processing industries dependent on the resources are looked at askance, as sunset industries whose day is fast fading. All too often, community, regional, and/or national energies are directed primarily at how to promote the supposedly more glamorous manufacturing and service industries. The resource sectors are considered to be dinosaurs, relics of a bygone age.[2]

This perspective is discussed in the next section. But before doing so, let us consider one or two other characteristics of the globalization process.

Another aspect of the globalized nature of production is the variety of policies which countries, in spite of the moves to greater trade liberalization, have been using to further the interests of their own corporations, and through these corporations, their own national well-being. These measures are usually included under the rubric, or euphemism, of "strategic trade policies"—which sounds so much more sophisticated and respectable than the word "protectionism". Such policies have been classified into two types: (1) profit-shifting strategies, and (2) learning curve strategies.

The first type of strategy is designed for world markets where relatively few firms exist so that each firm has some degree of monopoly power. The government gives its domestic champion a subsidy so as to strengthen its ability to underprice its competitors and force them to take a smaller share of the world market. The idea is that the economic clout of the corporation is augmented by the government assistance so as to overwhelm foreign competitors not having the same degree of government support. In this way, if the strategy is successful, the corporation, through its larger market share, is enabled to earn sufficient extra profits to more than compensate for the subsidy—thereby benefiting the nation as a whole.

The second strategy entails restricting access to the domestic market via the usual techniques of tariffs and quotas, or a variety of more subtle measures giving preference to domestic suppliers, so that a domestic firm or several domestic firms, through having sole access to the national market, can get well established, achieve economies of scale, and move well down its or their learning curve. The objective is to develop the domestic industry so that it can eventually compete successfully in the export markets of the world. That is, initial protection against imports is used as the way to facilitate subsequent exports.

In each of these strategies, government assistance in one form or another is central to the promotion of domestic industry and the resulting beneficial effects on employment and incomes. In the last four or five decades nations like Japan and South Korea have undoubtedly been the most successful in using these types of policies to advance their perspective places in the world community. But all the developed nations of today have, over the years, used one variant or another of these strategies to strengthen their economies. The interesting thing is to see the ardour with which they now want to prevent other countries, still at much earlier phases of development, from utilizing the same policies for their economic advancement.

In fact, what we are witnessing is the setting into place of rules restricting international government and corporate behaviour, rules that are being designed primarily by the dominant industrialized nations of the world—the United States, Japan and the EU—in conjunction with, and for the benefit of, their own domestic corporations.

A third significant dimension of the globalization process, which was referred to only in passing when we initially defined globalization, is that consumer tastes and standards are becoming increasingly international as a consequence of the globalization process in which the means of rapid communication are expanding exponentially. Common values with regard to consumer products and services—clothing, food, music and other entertainment, transportation systems, and so on—are arising in many different lands, even though many people are not yet able to possess all the things to which they aspire. What is really involved in many instances is that a common culture is developing because of the advances in communication and transportation. Some tightly-knit societies, such as the Japanese society, currently are able, and probably will continue to be able, to preserve much of their traditional culture and their ways of conducting business and government. But even there, many aspects of their lives are being swayed by the internationalization process.

What is fairly obvious is that the United States, because of its enormous economic size, the magnitude and economic clout of its entertainment and communication industries in particular, its attitudes and view of its own destiny in the world, and its unabashed willingness to use its economic power to further the international ends of its corporations, is often the source of much of the standardization of thought and preference among consumers, business and governments. There is often an attitude emanating from the United States that if things are not done the American way, they are unacceptable or wrong.

Canada, because of its proximity to and extraordinary economic dependence upon the United States, is particularly vulnerable to this United States influence. What we do in this country, what we are, how we see ourselves, is frequently inspired, consciously or unconsciously, by what the United States does, or is, or believes. This is true whether it has to do with the size of stumpage charges on our timber, our use of the Canadian Wheat Board to market our wheat internationally, our regional development programs, the television programs and movies we watch, or the magazines we read. For Canada, to a large degree, certainly to a greater extent than for other nations, globalization and internationalization means Americanization.

These, then, are the major characteristics of the globalization process of relevance in this volume. Having delineated them, the next step is to reflect briefly on them from an Innisian perspective, and relate them to the forestry industry and associated Canadian issues. The following section, therefore, considers Canada's trade in the light of the global tendency to focus on high-tech manufacturing and service industries, to the relative neglect of the resource-producing industries and the processing industries dependent upon resource inputs. In the subsequent sections, strategic trade policy and Americanization issues will be considered.

GLOBALIZATION, INNIS AND CANADIAN MERCHANDISE TRADE FLOWS

The Innis Perspective

The reader will recall that Harold Innis is best known for what has been labelled the "Staples Theory of Growth". The essence of this theory is that Canada's economic development and income per capita gains depended on the successful exploitation for export of a series of staple products beginning with fish, furs, and timber, followed by grain and a variety of metals and minerals. The benefits from the expansion of these industries were increased as backward linkages helped to stimulate the production of capital goods industries and other inputs, forward linkages increased the processing of the resources domestically prior to export, and final demand linkages from the incomes and spending power generated in Canada produced additional new investment to satisfy this internal demand (Watkins, 1963).

Given the discussion of globalization above, the temptation exists to see this staples theory, emphasizing, as it does, natural resource production and processing, as relevant only to Canada's past development, not in touch with present conditions, and certainly irrelevant for the nation's future economic advance. Yet, two types of data and analysis are available which strongly suggest that the Innis perspective is not as outdated as one might think. The first of these is the most recent trade statistics, and the second is two studies by the Economic and Trade Policy unit of Foreign Affairs and International Trade. Consider these in turn.

Canada's Merchandise Trade

The merchandise trade statistics for Canada and British Columbia are summarized in Tables 8.1, 8.2, and 8.3. Several observations are worth making. First, for 1994, natural resources and resource-based processed products still accounted for nearly 50 percent of Canadian commodity exports. If the huge automotive sector is excluded, which has been built up as a consequence of two special deals between the Canadian and United States governments, and the Canadian government and the automobile companies respectively (made about 31 years ago), resource-based products comprise two-thirds of Canadian exports (Table 8.1).

Second, the forest products sector was the largest single resource-products group, amounting to $31 billion for 1994, or nearly 14 percent of total Canadian exports over this period (18.5 percent if automobiles and parts are excluded from the total) (Table 8.1).

Third, for British Columbia, resource products are of much greater importance, accounting for 90 percent of total exports. Forest products alone were over three-fifths of provincial merchandise sales abroad (Table 8.3).

Fourth, although some of the resource sectors have had some difficult years of late, for 1994 the improvement in either prices and/or volumes was such for a number of the resource industries that the growth rates in their export values have equalled or even outpaced the rates for the remaining industries (Table 8.1, Column 3). They are not fading into the sunset.

Fifth, the resource-based products provided a $50 billion trade surplus for Canada for 1994, whereas the remaining industries ran a $25 billion deficit over this time. And this deficit would have been $10.6 billion larger were it not for the surplus of this size in the automotive sector (Tables 8.1 and 8.2)

Finally, for British Columbia, resources produced a surplus of over $14 billion; more than enough to offset the deficit of $10.3 billion on trade in other products.

In brief, where would Canada and British Columbia be in world trade at this time without the resource industries? They remain a large and vital part of our domestic economy and international relationships.

The Domestic Impact of Exports

Of course, trade is not the whole story. The modern emphasis on globalization is also concerned with the effects of global adjustments on domestic employment and incomes. As noted, such concern is also a vital aspect of Innis' theory of the importance of staples, or resource production. This is where the work by the Economic and Trade Policy staff of Foreign Affairs and International Trade is of considerable significance.

The first study in this regard (Martin, 1993) showed how incorrect it is to think that because merchandise exports account for roughly 25 percent of GDP,

Table 8.1   The commodity composition of Canadian merchandise exports: January-December 1994

Product Groups	$ Billions (1)	(2)	% (3)	Growth (%) Jan-Dec 1994 (4)
RESOURCE-BASED PRODUCTS				
Agricultural Products		14.65	6.5	14.2
Fish and Fish Products		2.95	1.3	10.6
Energy Products				
Crude Petroleum	7.43			7.6
Natural Gas	6.85			18.3
Coal	2.12			11.3
Petrol. & Coal Products	4.00			3.6
Electricity	1.31			53.0
Total Energy		21.71	9.6	12.4
Forestry Products				
Lumber & Saw Mill Products	14.06			24.8
Wood Pulp & Other Wood Products	6.72			35.9
Newsprint, Paper & Paperboard	10.37			6.7
Total Forestry		31.15	13.7	20.7
Industrial Goods and Materials				
Metal Ores	3.67			6.1
Metals and Alloys	15.89			20.6
Chemicals, Plastics & Fertilizers	11.54			26.1
Other Industrial Goods & Materials	8.31			24.2
Total Industrial Material		39.41	17.4	21.4
Total Resource-Based Products		**109.87**	**48.5**	
OTHER PRODUCTS				
Machinery and Equipment		43.05	18.9	27.3
Automobiles and Parts		58.45	25.5	20.5
Other Consumer Goods		5.83	2.6	23.6
Special Transactions, Trade		9.43	4.1	30.5
Total Other Products		**116.76**	**51.1**	
TOTAL		**226.63**	**100.0**	**21.0**
Resource-Based Products as a Percentage of Total Trade Excluding the Automotive Sector			**65.6**	

Source: Statistics Canada. *Canadian International Merchandise Trade. Dec. 1994.* Ottawa, 1995.

Table 8.2   The commodity composition of Canadian merchandise imports: January-December 1994

Product Groups	Imports $ Billions (1)	% (2)	Growth (%) Jan-Dec 1994 (3)
RESOURCE BASED PRODUCTS			
Agricultural Products	11.40	5.6	14.0
Fish and Fish Products	1.17	.6	17.0
Energy Products	7.11	3.5	4.6
Forestry Products	1.81	.9	15.5
Metal Ores, Metals, and Alloys	9.63	4.8	25.2
Chemicals, Plastics and Fertilizers	13.72	6.8	23.6
Other Industrial Goods and Materials	15.35	7.6	18.4
Total Resource-Based Products	**60.19**	**29.8**	**17.7**
OTHER PRODUCTS			
Machinery and Equipment (excluding Automobiles and Parts)	65.58	32.5	23.6
Automobiles and Parts	47.82	23.6	19.4
Other Consumer Goods	23.51	11.6	10.0
Special Transactions - Trade	4.93	2.4	13.3
Total Other Products	**141.84**	**70.2**	
TOTAL	**202.02**	**100.0**	**18.9**
Resource Based Products as Percentage of Total Trade Excluding the Automotive Sector		**39.0**	

Source: Statistics Canada. *Canadian International Merchandise Trade: Oct. 1994.* Ottawa, 1994.

they also account for about 25 percent of domestic employment. (This is a mistake which many people continue to make in this country.) The research showed that, counting not only the direct jobs going into such exports but also the indirect service inputs entailed, merchandise shipments abroad were responsible for only 13.4 percent of total Canadian jobs. This occurred because many of the exported "end" products—such as machinery and equipment, automobiles, and consumer goods—have substantial proportions of imported inputs. Overall, import content comprises about 25 percent of the value of *all* Canadian merchandise exports, and for the sectors just mentioned, this proportion is much higher. In contrast, resource-based products have relatively little foreign content in their output—less than 6 percent[3]—and so account for a larger share of employment than their dollar value would suggest.

These variations in import content and related characteristics of exported goods became the subject of detailed investigation in the second Trade Policy

Table 8.3   Merchandise trade of British Columbia: January-December 1994[1]

	Exports			Imports		
	$ (Billions)	% of Total B.C. Exports	% of Total Canadian Exports	$ (Billions)	% of Total B.C. Imports	% of Total Canadian Imports
Product Group	(1)	(2)	(3)	(4)	(5)	(6)
RESOURCE BASED PRODUCTS						
Agriculture and Fishing	1.53	6.7	8.7	2.16	11.9	17.2
Energy	2.20	9.6	10.1	.22	1.2	3.1
Forestry	13.99	61.3	44.9	.43	2.4	23.8
Industrial Materials	2.74	12.0	7.0	2.77	15.2	7.2
Total Resource Based Products	**20.46**	**89.7**	**18.6**	**5.58**	**30.8**	**9.3**
OTHER PRODUCTS						
Machinery and Equipment	1.49	6.5	3.5	5.72	31.4	8.7
Automotive Products	.37	1.6	.6	3.78	20.8	7.9
Consumer Goods	.30	1.3	5.1	2.66	14.6	11.3
Special Transactions - Trade	.18	.8	1.9	.47	2.6	9.5
Total Other Products	**2.34**	**10.3**	**2.0**	**12.63**	**69.4**	**8.9**
TOTAL	**22.81**	**100.0**	**10.1**	**18.19**	**100.0**	**9.0**

Source: Statistics Canada. *Canadian International Merchandise Trade, Dec. 1994.* Ottawa, 1994.

[1] It is recognized that these numbers are subject to correction. Statistics Canada has no straightforward, or completely accurate, way of tabulating the proportion of exports leaving a province which came entirely from that province, or the proportion of imports actually consumed in that province as opposed to other parts of Canada. But the numbers are certainly sufficiently accurate to draw the broad conclusions we have drawn in the text.

study (McCormack, 1994). In this recently released paper, an attempt was made to identify those exporting industries which contribute the greatest benefits to the domestic economy. It was deemed that those export industries which: a) had a high exports to industry output ratio; b) relied the least on imported intermediate inputs, and correspondingly; c) involved a very significant domestic value-added as a proportion of export value, and whose; d) production entailed a substantial number of jobs in relation to domestic value-added, and/ or; e) jobs with higher than average salaries, were the ones that contributed the most to the national economy.

The study was careful to point out that it was not meant to suggest either that a nation should eschew those internationally integrated industries which, because of the globalization process, involve a high proportion of imported inputs, or that import substitution policies regarding such inputs should be put into effect for these industries. Its main focus was to determine "what is" (that is, positive economic analysis), not to say "what ought to be" (normative economics).

The results provide a useful counterbalance to the constant stream of verbiage we hear and read about the urgency of promoting the high-tech industries, and that our national survival depends almost solely upon our success in this regard. *The natural resource producing industries*, particularly agriculture and related services, mining, and crude petroleum and natural gas, in general produce the greatest benefits as judged by the five criteria set out above.[4] The other group of industries which had an above average performance in terms of the domestic output, employment and income benefits produced, are *those manufacturing industries exporting processed natural resources* such as wood and paper products, and food and beverage products.

The wood industries, especially the largest exporting segment—sawmills, planing and shingle mills—performed the best of all resource and processing industries, satisfying all five criteria[5] (Ibid.: 36-39).   That is, they had an above average exports to output ratio, a below average reliance on imported intermediate inputs, and significant domestic value-added. They also created an above average number of jobs in relation to domestic value-added, and paid above average salaries.

The remaining manufacturing industries, including those classified as high-tech industries, generally did not perform as well. Some, for example, produced high-paying jobs, but not large numbers of them in comparison with domestic value added. Others created a significant number of jobs, but not ones that paid well. The high-tech industries—aircraft and aircraft parts, telecommunication products, electronics parts and components, computers and related products, miscellaneous office and business machines, pharmaceutical and medicines, indicating and recording instruments and other scientific and professional equipment—as well as the automotive sector, all tended to depend highly on intermediate inputs from abroad. In other words, the backward linkages to domestic production in the form of technology transfers with new supplier industries being developed—although frequently the stated reason for promoting these high-tech industries—are not much in evidence at this time. The high-tech industries, because their intermediate inputs are frequently from abroad and their output is mostly exported, tend to function, in some ways, almost in isolation from the balance of the economy (Ibid.: 56). And as McCormack says: "In terms of *measurable* domestic economic gains, such as increases in employment or GDP, high-tech industries are not an obvious choice for the targeting of export initiatives." (Ibid.: 4)

The research paper brings this point home in another way, too—by estimating the Canadian value-added implications of our well-known merchandise trade surplus with the United States. For 1993 this surplus amounted to $19.7 billion. Because Canada's exports to the United States involve a larger proportion of manufactured goods than is true for exports to other markets, there is a tendency to look with uncritical favour upon this trade balance. The study's results tend to dampen somewhat any euphoria we may have about this trade surplus. First it calculates the import content in the various types of exports to the United States (24.7 percent for manufactured goods taken as a group—although the percentage is about double this for the large transportation equipment industries like automobiles and aircraft and electrical and electronic products industries), and then it estimates the Canadian content in goods imported from the United States (in the neighbourhood of 12 percent). The net result of these adjustments is that what was nearly a $20 billion trade surplus with the United States (on a balance of payments basis) turns into a value-added *deficit* for Canada of $4 billion!

Although one might wish to quibble with some of the assumptions necessary to arrive at this particular number, different assumptions would not alter the overall conclusion. The basic message given is valid and worth noting; namely, that we should look beyond mere aggregate trade balance figures when assessing the domestic benefits from trade with other lands, and consider the net effects on the domestic economy. And when we do that vis-a-vis our largest trading partner, accounting for 82 percent of our exports, the net gains to the domestic economy of such trade are considerably less than a quick glance at the trade statistics would suggest. This occurs because the high-tech and other manufactured goods exports, like automobiles, contain substantially less value-added than do resource-based products.

Concluding Comments

In short, all that has been said in this section suggests that natural resource endowments, and particularly forest endowments, will continue to have great importance for the Canadian economy. We should clearly not be neglecting the wise management of our primary and secondary resource-based industries in the globalization process occurring today. They currently contribute much to our economy and have the potential to continue to do so in the years ahead.

Of course, a vital aspect of this management process involves the continuous appropriation and application of pertinent technologies and skills from the high-tech or knowledge industries. Only in this way can the resource-based sectors continue to be efficient, internationally competitive, and prosperous. In this regard, too, the resource sectors in many instances already have their own high-tech/knowledge-based segments. Hence, it need not, indeed should not, be a matter of competition between the two types of sectors, but one of complementarity. Innis would have no difficulty with this prospect, for, on the one hand, he continually noted that cooperation and complementarity among sectors was one of the keys to continued economic progress for the nation as a whole, and on the other hand, emphasized "the importance of technology as a factor in economic change" (Creighton, 1957: 102) in his study of the staples industries.

STRATEGIC TRADE POLICY IN A CANADIAN CONTEXT

Innis would have been readily able to relate his perspectives to the strategic trade policies of this age. He noted, and endorsed the need for, government assistance in the development of the resource industries, whether that support took the form of direct government involvement in the building of railways or canals to enable resource products to get from the hinterland to world markets, or government lending or guaranteeing the debts of corporations trying to build these transportation links or exploit the resources. He saw some of these measures not simply as isolated policies to advance the Canadian economy per se, but as essential if the progress already made in the economy was not to be eroded by the rapid investment and general progress in the United States. It was a matter of keeping up (Neill, 1972: 62-63). Government capabilities were mobilized to assist where the private sector was having difficulty financing the expansion required for competitive purposes—not unlike the philosophy of strategic trade policy of today. The main difference is that then the focus was mainly on the exploitation of natural resources, whereas today the focus is mostly on sophisticated manufacturing and knowledge-based industries.

Another related aspect of Canada's situation in the global economy today which Innis would be very much in tune with is as follows. It is not a matter for detailed discussion in this volume, but it is at least worth mentioning briefly in that how it is dealt with certainly will affect the future international competitiveness and prosperity of the entire forestry sector. Innis had great concerns about it in his day, and we have even more apprehension about it today. It is the massiveness of Canada's debts, in particular its indebtedness to non-residents, and the burden of servicing and repaying this debt. The origins of the problem are different in the two eras however. When Innis was writing,[6] he saw the debt as stemming primarily from the activities of government in helping to finance the development of the staples industries, and the servicing problem arising because the price of staple products had collapsed in the early 1930s. In our day, although government grants and privileges to industry have certainly contributed to the size of the debt, it is probably fair to say that the basic problem has arisen because of profligate expenditures by successive Canadian governments trying to bribe voters with what was essentially their own money.

No more need, or should, be said about the issue at this juncture. It is sufficient just to observe that it is a problem which Innis would have been one of the first to detect and express dismay about.

GLOBALIZATION AND THE AMERICANIZATION OF BUSINESS, COMMUNICATIONS AND CULTURE

After Innis and his associates had virtually finished the work on Canada's traditional staple industries, he turned his attention to the communications sector—to a study of its history and development, and its implications for the evolution of ideas, cultures and nations. He saw communications as being at the core of the economy—as indeed they are today. He noted, as well, how technical developments in communications have, over the centuries, intensified the monopolistic influence of the metropolis (Mackintosh, 1953: 2)—a phenomenon which we certainly see in the pre-eminence of the Londons, New Yorks, and Tokyos of the world. His work also made clear the usefulness of researching communications advances in other lands so as to provide a perspective from which to view, and begin to understand, actual and potential developments in his own land.

Throughout his academic career, but particularly in its later stages, Innis was sensitive to and concerned about the dominant role the United States played in Canada's life. For Innis, these two topics—communications, particularly the monopoly power that can be created, and the Americanization of Canada —went together.

This unity is best seen in one of his last publications, *The Strategy of Culture* (1952). In the first essay of this little volume Innis noted how the United States, in stark contrast to today when it sees copyright violation almost as one of the seven deadly sins, used copyright violation of British publications and highly protective measures to develop its own domestic publishing industry.[7] He observed, too, that as technological change occurred, it "led to the creation of vast monopolies of communication" (p. 15). For Canada, he saw these as resulting in a "continuous, systematic, ruthless destruction of elements of permanence essential to cultural activity" (p. 15). The accessibility of United States radio and television, and United States periodicals with their "numerous suggestions . . . that Canada should join the United States" (p. 19) were seen by him as threatening the continuation of a separate Canadian cultural life and identity.

In his earlier writings he had of course seen the great importance of the United States in Canada's initial development. Investment in canals and other transportation facilities in Central Canada were noted to be essential if Canada was even to begin to match the expansion to the south. And the transcontinental railway was a means not just of bringing resources from the Canadian interior to world markets, but of tying the nation together against aggressive United States aspirations to control all of the continent. But in his later years he became very concerned about the naivete of those who saw Canada's main threat as being from British Imperialism, who saw the reduction of Canada's dependence upon Britain as the key struggle in achieving nationhood, and who ignored or minimized that in the process of bringing about this separation from Britain, Canada was increasingly, and often unwittingly, falling into the vortex of United States economic and cultural life (Creighton, 1957: 103-105). By the time of his *Strategy of Culture* essays, his ideas in this regard were pretty strongly developed. One quote will demonstrate this:

> *"We are indeed fighting for our lives. The pernicious influence of American advertising reflected especially in the periodical press and the powerful persistent impact of commercialism have been evident in all the ramifications of Canadian life . . . We can only survive by taking persistent action at strategic points against American imperialism in all its attractive guises"* (pp. 19-20).

The evident dismay conveyed by these words was merely the culmination of the unease he had had for some time, as outlined in a letter written in late 1949 commenting to a friend on his time in Ottawa. He saw Lester Pearson as "selling us down the river" (cited in Creighton, 1957: 133), and said ". . . one feels how much we have lost to the United States. We follow along and kid ourselves we are our own masters" (Ibid.: 138).

In the second and final essay of his *Strategy of Culture* volume, "The Military Implications of the American Constitution," Innis, among other things, noted American aspirations to dominate this hemisphere. He saw this evidenced in the 1823 Munroe Doctrine "which warned European powers to keep their hands off South America" (p. 22), in the "absorption of Texas, California and other states at the expense of the Spanish empire and of Mexico" (Ibid.), and in such rallying cries as the one by Henry Cabot Lodge in 1898 that "From the Rio Grande to the Arctic Ocean there should be but one flag and one country" (p. 33). He concluded his essay with this comment:

> *"Ostrogorski has quoted the remark that God looks after little children, drunken men, and the United States. I hope it will not be thought blasphemous if I express the wish that He take an occasional glance in the direction of the rest of us"* (p. 45).

Some Canadians, as soon as sentiments such as these by Innis are uttered, dismiss them as "conspiracy theory." This label is supposed to be sufficiently pejorative in nature that no further discussion is considered necessary. But what the users of this term really seem to be saying by this unwillingness to investigate the issues is that they simply have no answers and really do not want to discuss the matter for that reason. Their implicit hope seems to be that it will go away. I doubt it will be that simple. Innis had a realistic feel for the situation.

In this regard, the forestry sector has certainly experienced its share of US aggression and desire to control. Many of the readers of this volume will be all too familiar with, and probably very weary of, the softwood lumber disputes with the United States under Chapter 19 of the Canada-United States Free Trade Agreement and/or NAFTA, as well as the extraordinary challenges, and other machinations of the United States political economy processes.

This type of United States action should not be seen as merely a temporary US response to the globalization processes we have been examining, and the uncertainty for and threats to American interests which these processes are creating. The historical analysis of United States behaviour which Innis and a considerable variety of other eminent Canadian historians, such as Donald Creighton (e.g. 1956) and Hugh Keenleyside (1929) have done over the years, suggests, that the recent aggressive use of US contingency policies, with regard to lumber, fish, agricultural products or other Canadian staples, represents a continuation of a long-standing United States policy of using whatever means seem suitable at the time for furthering its interests, regardless of the implications for the sovereignty or well-being of other nations, such as Canada. It is one of the facts of our North American life. The globalization forces seem only to have increased the intensity of some of these American responses.

The main implication of this issue for the reader of this volume is that, as other concerns regarding the future of forestry are being discussed, it is important always to be looking at them not from a narrow domestic perspective, but from the viewpoint of both the globalization processes now under way, the United States response to these processes, and the long-standing US conception of its own destiny and interests. In this endeavour, we need constantly to recognize the interrelationships between communications, industrial development, economics, politics and culture. They are not isolated considerations.

This brings me to my final observation on globalization, Innis, and the forestry sector. It is of sufficient significance that it deserves a short, concluding section unto itself.

INNIS, ECONOMIC MODELS, AND MONOPOLIES

We have already noted that Innis decried the strong monopolistic dimension of the United States cultural industries. But he had no use for another type of monopoly either—the monopolies that arise in academic theorizing and investigation. He saw these as stifling thought, narrowing the field of investigation, and hindering progress. He often chose, based on his detailed analysis of the facts, to challenge accepted views, and warned his students to be prepared to do the same (Creighton, 1957: 129-130).

We need to have the same questioning attitude today with regard to various aspects of the globalization-Americanization process taking place and to some of the commonly accepted views regarding the appropriate response thereto. For example, the popular, ruthless view that suggests workers are just another input, to be treated with no more respect or consideration than a piece of obsolete or worn out capital, needs to be questioned. So must the view that the market, in itself, will look after all societal interests, be they environmental, distributional, the preservation of replenishable resources, or whatever.[8] We could go on, but enough has been said to affirm that we would be wise to retain that independence of thought about changing circumstances that characterized Innis' work.

Also of much relevance for us today in this regard is to note that Innis did not develop his views sitting in an ivory tower. Insofar as they related to Canada, he derived them not only from a meticulous examination of the historical record, but from travelling the length and breadth of the land, getting a first-hand feel for the people and their circumstances. In addition, although he saw much benefit from studying the history of other lands and eras, he did not ever take theories or models from abroad and simply try to squeeze them into a Canadian mould—as seems so often to have been done by academics, or others who have received their graduate education in other nations, particularly in the United States which dominates in this area as in every other area of Canadian life. He developed his views based on his own analysis of Canadian conditions. Except for his international stature, he might not have been as popular in Canada. And certainly, today, his willingness to challenge "accepted" monopolies of thought, would probably have made him rather unpopular among many groups both inside and outside academia. But the very fact that this book is being published in honour of the 100th anniversary of his birth, indicates that his concern for the well-being of our nation and his way of expressing that concern have had a worthwhile and lasting impact on our land.

ENDNOTES

[1]   In saying this, I am not attributing this view to the authors cited above. I am simply saying that all too often our society more generally leaps to this conclusion - and as I will show below, those who do so, overlook some important facts and considerations.

[2]   If my readers think I am being a bit too strong in my use of words here, they might refer to the article "Old Economy Roaring Back" in *Globe and Mail* of January 27, 1995, p. B-1 where the word "dinosaur" is used for the resource industries.

[3]   This percentage is actually from the second study discussed in the subsequent paragraphs by McCormack, p. 34.

[4]   They were above average on four of the five criteria employed. Agriculture and related services fell short on one criteria, namely low average wages. This was, in part, because the incomes of this sector "are part of the unincorporated business income and `other

surplus' in the national accounts" and are accordingly difficult to adjust to enable accurate comparison with other sectors (McCormack, 1994:37). The mining and crude petroleum and natural gas industries were strong performers on four criteria, but mere below average in the number of jobs they "created per dollar of GDP in the production of exports" (Ibid.).

5   The other sector which, domestically, performed well was Educational Services. These industries are also important in that they provide value-added support to exporting industries. But they are not in themselves export-oriented, and are not the focus of this particular paper.

6   Such as in the Preface to *Problems of Staple Production in Canada*.

7   The American Copyright Act of 1790 gave *no* protection to British books, and indeed encouraged United States firms to publish the writings of British authors without having to give royalties. The Copyright Act of 1891 provided that for books to receive copyright protection, they had to be published in the United States (Innis, 1952: 3-4).

8   For example, will the market produce the appropriate balance between forestry and fish spawning grounds? Or will a market dominated by Cargill, the largest private corporation in the world, which controls 25 percent of the entire world's grain trade, be a superior marketing agent for Canadian wheat farmers compared with the Canadian Wheat Board, which has farmers' interests at heart?

REFERENCES

Courchene, T. (1994). *Social Canada in the millennium: Reform, imperatives, and restructuring principles*. Toronto: C.D. Howe Institute.

Creighton, D. (1957). *Harold Adams Innis: Portrait of a scholar*. Toronto: University of Toronto Press.

_____ (1956). *The Empire of the St. Lawrence*. Toronto: Macmillan.

Drucker, P.F. (1994). The age of social transformation. *The Atlantic Monthly*, November, 53-80.

Globe and Mail (1995). January 27, p. B-1.

Harris, R. (1993). Globalization, trade and income. *Canadian Journal of Economics*, XXVI(4), 755-776.

Hart, M. (1993). A brave new world: Trade policy and globalization. *Policy Options*, 13(10), 3-7.

Innis, H.A. (1952). *The strategy of culture*. Toronto: University of Toronto Press.

_____ (1951). *The bias of communication*. Toronto: University of Toronto Press.

_____ (1950). *Empire and communications*. Toronto: University of Toronto Press.

_____ (1946). *Political economy in the modern state*. Toronto: The Ryerson Press.

_____ (1933). *Problems of staple production in Canada*. Toronto: The Ryerson Press.

_____ (1930). *The fur trade in Canada: An introduction to Canadian economic history*. New Haven.

Keenleyside, H. (1929). *Canada and the United States: Some aspects of history of the Republic and the Dominion*. New York: Knopf.

Mackintosh, W.A. (1953). Innis on Canadian economic development. *Journal of Political Economy*, LXI(3), 185-194.

Martin, M. (1993). *Exports and job creation*. Ottawa: Department of Foreign Affairs and International Trade, Economic and Trade Policy Division.

McCormack, J. (1994). *The impact of exports: An input-output analysis of Canadian trade*. Ottawa: Department of Foreign Affairs and International Trade, Economic and Trade Policy Division.

Neill, R. (1972). *A new theory of value: The Canadian economics of H.A. Innis*. Toronto: University of Toronto Press.

Watkins, M. (1963). A staple theory of economic growth. *Canadian Journal of Economics and Political Science*, XXIX(2), 141-158.

Plate 13 (overleaf)   Inside a sawmill   ➤

A Changing Global Context for British Columbia's Forest Industry[1]

Patricia Marchak

*Department of Anthropology and Sociology,
University of British Columbia*

Harold Innis would surely have viewed British Columbia between the 1880s and 1970s as the perfect example of a staples economy. Enjoying an embarrassment of resource riches, it failed to develop economic alternatives to export sales of semi-processed products (dimension lumber and pulp). A workforce was established for extraction and processing, but its upward end was truncated. Scarcely any research was done within the regions of extraction. Imported management and professional personnel followed policies determined in distant head offices, often located in the United States. The extraction industry was heavily subsidized by governments that formally owned the land but leased harvesting rights to large, integrated corporations. A handful of these controlled the larger part of all forest licenses and other logging rights. The practices and technology of the industry were destructive of wildlife habitat, the forest itself, streams, rivers, and ocean life, but government, companies, and organized labour maintained that no alternatives were economically viable (Marchak, 1983).

Change has occurred since 1980 with causes both local and global. Resource depletion and the time span required for replacement of northern forests are major causes of change at the local level. Employment relative to production volume has steadily decreased since the early 1960s, and employment has decreased absolutely since 1980. Accompanying these changes is a major change in the context of industrial forestry as new capacity emerges in southern climates. This chapter will concentrate on the global context, and will connect the context to British Columbia in a concluding section.

JAPAN'S SEARCH FOR WOOD

Forests cover a higher percentage of Japan's total territory than of any other country save Finland, and most of these forests were planted following 1945. But domestic reserves cannot supply the high-demand construction and paper-making industries. Japan is a country of wooden dwellings, and construction timber has always been a high demand item. Up to the mid-1970s Japan imported tropical hardwoods for plywood and construction from South East Asian countries, temperate-zone conifers from Russia and the United States for construction. Through the 1970s, South East Asian countries, one by one, either ran out of wood or imposed log export bans; Indonesia still had wood but developed its own plywood industry and has since become the world's top supplier of tropical plywood. The Philippines, Thailand, and mainland Malaysia stopped exports, but Malaysia allowed continued exports on Sabah and Sarawak; they and the teak forests of Myanmar are still being logged for export. With declining tropical sources, Japan became more dependent on North American, Russian, and New Zealand conifer forests for construction wood.

Ironically, North American logs of high quality are sold on the log markets of Japan at lower prices than domestic logs of similar quality. Japanese forest economists attribute this to the undervaluing of standing timber in North America, the lack of obligation to replant in like quality or to engage in intensive management similar to that undertaken and legally binding for forest owners in Japan (Marchak, 1992). Other economists argue that the cause is the inefficient structure of Japanese forest ownership arrangements and the sawmill industry together with high wages and low labour availability in rural regions of Japan (Cox, 1988). With affluence, the Japanese housing industry has continued to expand so that increasing quantities of lumber are also being imported from North America and New Zealand.

The Japanese paper-making industry is even more voracious in its search for wood sources than the construction industry. Japan is second only to the United States as producer of paper products and virtually all its produce is sold on the domestic market. There are many pulp, paper, and specialized paper mills on the islands. Half of their raw material consists of wastepaper and non-wood fibres, but they seek wood fibre for the remainder and indigenous sources are insufficient. Tropical woods are fed into the pulping machines, but there are few remaining sources in South East Asia. Thus have Japanese companies become globetrotters in search of new fibre sources

In 1987, 43 percent of total pulpwood supply in Japan was imported; 59 percent by 1993. The difference reflects the growth in the domestic market for paper products more than any diminution of domestic wood sources. Over the 1987-1993 period, the United States annually provided between 5.5 and 9.2 million cubic metres of wood, slightly over half of this in softwood. Canada provided between 1 and 2 million cubic metres, almost all softwood (Japan Paper Association, 1994). Canadian log restrictions prohibit higher exports. British Columbia has always maintained export restrictions, hovering around 2.4 percent of the total harvest except during the early 1980s when exports were permitted to rise to above 4 percent of the harvest. In the United States, however, private land-holders have been able to export as they pleased even though exports of timber culled from national and state forests are restricted. As long as the Yen and other conditions provide higher prices for logs in Japan than for lumber or pulp in the United States, a substantial export trade from the US Pacific states to Japan will continue. It was reported to be about 25 percent of the total harvest from private lands in Washington and Oregon states over the late 1980s and early 1990s. Weyerhaeuser is the major exporting company.

Where sufficient supplies of logs were unavailable, Japanese companies turned to wood chips. A fleet of specialized wood chip carriers transported over 80 percent of the world's wood chip supply to Japan from the 1960s through the early 1980s. Weyerhaeuser was the largest single supplier in the 1970s, and the western United States and British Columbia were the major supply regions to Japan. In 1979 United States and Canadian suppliers upped their price by 67 percent over a six month period. Japanese buyers went into "chipshock" and Japanese policy thereafter included expansion of domestic pulpwood supplies, increase in the wastepaper component, and diversification of overseas fibres supply sources. Among the new suppliers were Chile and Australia. The chip trade tripled in the early 1980s but more countries are engaged in the market (Schreuder and Anderson, 1988: 162).

As well, Japanese companies began investing in plantations and chip-production facilities elsewhere so that they were less dependent on an open chip market. Their major investments include two large plantations in Chile, one through Mitsubishi, the other through Daio together with C. Itoh. These are merchant companies (sogo shosha) buying on behalf of paper companies in Japan. The same two companies have shares in eucalyptus chipping operations in Australia, California and Oregon. Honshu, Jujo, Sanyo-Kokusaku, and Oji directly hold shares in other operations in Papua New Guinea, Indonesia, and British Columbia.

Just as Japanese buyers began to invest in plantations for wood chips rather than rely on volatile markets, they have also begun to invest in more extensive pulpwood plantations connected to pulpmills, or to free-standing pulpmills reliant on remaining indigenous forests in North America and Russia. Among these are the two Daishowa-Marubeni mills at Quesnel and a more recent one on the Peace River. But the largest of the new investments are in the boreal region of Alberta. The enormous AlPAC mill on the Athabaska River, 45 percent owned by Mitsubishi with minority shares held by Canadian shareholders and three other Japanese investment groups, received subsidies for mill construction amounting to some $275 million in addition to infrastructure and other expenditures from public funds on behalf of the mill. It has a 1,500 ton daily bleached kraft pulp production, that is over 500,000 tons annually and it will consume 3.2 million cubic metres of timber per year (Pratt and Urquhart, 1993; McInnis, 1994; Richardson, Sherman, and Gismondi, 1993). Japanese-owned pulp mills are now also located in Washington, Georgia, Massachusetts, and Alaska. A particularly large plantation pulp mill complex is Cellulose Nipo-Brasileira (Cenibra) in Brazil. Still others are in New Zealand, Thailand and Portugal.

TROPICAL DEFORESTATION, PLANTATIONS, INDUSTRIAL DEVELOPMENT

Forestry is not entirely new to the southern hemisphere: decorative woods were logged by imperial foresters in Africa and Asia from the beginning of European colonization. More recently the araucaria forest all along the Atlantic coast of Brazil was logged to produce ordinary lumber, and much of the wood was wasted in the process. Today tropical forests in Southeast Asia and Papua New-Guinea are logged for wood destined to become plywood or moulds for construction. But of all these industrial logging operations, wasteful and destructive as they were and are, none approached the levels of systematic deforestation for mass production of wood products that the northern industries established even before the turn of the last century. Industrial forestry of the kind that turned out dimensional lumber, pulp, newsprint, and paper was a northern industry until about the 1970s. Even as late as 1980 the future of a southern plantation-based industry was not assured. In the decade of the 1980s, however, the economic viability of plantations was demonstrated, especially in Brazil; and advocates for plantation forestry gained strength in international forums even while the products gained global market shares.

The Tropical Forestry Action Plan, devised in the mid-1980s under UN/FAO sponsorship, was supposed to reduce pressures on tropical forests by increasing the amount of wood products procured in plantations (FAO, 1985). Subsidies for planting and planning of crops and mills were provided to many developing countries with tropical forests. The plan was well under way when the Rio conference declared that all logging in tropical forests should cease as of the year 2000. In fact, tropical forests are being cut at high speed so that plantations can be installed before the turn of the century, and the land given to plantations is land that was forested. Deforestation is thus continuing but it is veiled by reforestation statistics, even though the new "forests" are not forests at all, they are agricultural crops.

Cloned, fast-growing trees—usually exotic to the region—are planted in precise rows to produce a homogeneous crop ideally suited to pulping. Pines are planted in temperate climates; *Pinus radiata* the most popular. It has been demonstrably successful in New Zealand plantations. Radiata and other pines together with hardwoods have been successful in the southern United States and Iberian Peninsula. Eucalyptus varieties, acacia, laeucanae, gmelina, and albizzia have been planted in warmer regions, with eucalyptus the most popular in Latin America, acacia the tree of choice in Indonesia. Eucalypts reach maturity within seven years, can be coppiced to produce three crops on a single stem, their produce per tree and per hectare is very high, and they produce high quality pulp suitable for writing, book, and computer papers.

There are known ecological defects to plantations. These crops do not create a multi-layered canopy, do not protect the soil, deplete the water-table, alter the micro-climate of the plantation region, may alter macro-climates as well, and cannot provide for either wildlife or cattle in the spaces between the plants. Nonetheless, they are relatively inexpensive to grow, produce cash crops in short time-spans, need very little infrastructure, and could provide developing countries with a major industry.

There are also severe social defects in many tropical and subtropical countries. Indigenous shifting cultivators lose their livelihoods and culture when the land their ancestors have travelled for centuries is transformed into plantations. Landless peasants also lose land and fuel sources as forests are destroyed to make way for plantations. The new plantation industry is supposed to provide employment, multiplier effects, and spawn new secondary industries in poor countries. The record to date shows governments and military juntas providing subsidies, land grants, infrastructure, and other incentives to plantation companies (Westoby, 1989). It is sparse on evidence about multiplier effects except in Brazil where employment and growth in secondary service industries is, indeed, evident.

Plantations in the southern United States produce pine softwood in 15 to 30 years, and these are now being used for newsprint production; as well, hardwoods in 10 to 30 years useful for other papers. Governments, Scandinavian and western European companies, even at least one American company, have shares in the plantations and pulpmills of Spain and Portugal. As Europe moves toward unification and economic barriers are removed, Nordic companies that used to supply Northern-grown pulp to paper companies in France and Germany are now major paper producing companies on the continent, drawing on Iberian wood sources. The southern United States and Iberian Peninsula were established plantation regions before 1970. Brazil and Chile became firmly established in the 1970s and 1980s. Indonesia, Thailand, Malaysia, and South Africa are all potential plantation regions with some investment already in place, and other countries are suitable but not yet organized for large-scale plantation development.

Brazilian Plantations

Most of the current industry is in the central-southern region, just inland from the Atlantic coast. This region is subtropical, with more temperate climates in a few mountainous areas. Clearing occurred for the most part before forestry plantations were viable, for other agricultural plantations or ranching, and indigenous peoples of this coastal region were already dispersed by the time pulp plantations became viable.

Unlike other developing countries, Brazilians have developed the plantation/pulp industry largely on private investment funds generated within the country, though they have been hospitable to foreign investment. Tax incentives and land grants were liberally dispensed by the Generals in the 1970s, and the National Bank for Economic Development (BNDE) became a major public support for development. BNDE includes a special agency for industrial financing (Finame) which has provided the funds for development of Brazilian-manufactured machinery and equipment for pulpmills. These enabled Brazilians to become owners in the industry, despite an earlier history shaped largely by external paper companies. Foreign companies are represented, but typically as secondary investors.

By 1985 Brazil had become the world's seventh largest pulp producer. Much of the output went into domestic markets which were growing as fast as capacity in the early 1980s, but since then virtually every mill has doubled or tripled its capacity and exports have increased. Pulp is exported primarily to Japanese paper manufacturers, while paper production is still destined mainly for domestic consumption.

The major companies are Aracruz; Klabin and its subsidiary, Riocell; Cenibra, Ripassa, Impacel, and Imprensa. Aracruz is in the state of Espirito Santo, the largest single forestry project in Brazil and the world's leading producer and exporter of bleached eucalyptus pulp. About 122,000 hectares are now planted or planned for eucalypts, with a planned rotation age of seven to eight years. The mill produces over a million tons per year—about 20 percent of the world's total supply of kraft sulphate pulp—and exports about 80 percent of this. Aracruz is the model for developments in Asia.

Cenibra, located in the state of Minas Gerais on the Rio Doce River, is a joint venture between a consortium of 18 producers of paper and pulp in Japan, the Japanese government's Overseas Economic Cooperation Fund, a trading company, C. Itoh, and the Brazilian government iron ore mining company Campanhia Vale do Rio Doce (CORD). The Japanese consortium includes Oji, Honshu, Daishowa, and Mitsubishi. Cenibra owns over 145,000 hectares of forest land, and had planted some 85,000 by 1992. Klabin, one of the oldest pulp mill companies in Brazil, owns 18 plants including its major paper and newsprint mills at Monte Alegre in the state of Parana, and Riocell at Porte Alegre in the state of Rio Grande do Sol. It is ranked 57th among major pulp and paper companies by Pulp and Paper International (1990), a ranking that is higher than most Canadian paper companies in the 1990s. Klabin has nearly 300,000 hectares, two-thirds of this in radiata and loblolly pine, eucalyptus and araucaria, the remainder in native forests. Ripasa, an integrated company owned by Brazilian families and situated at Limeira in the state of Sao Paulo, includes five large mills resting on 84,000 hectares in Sao Paulo and new plantations in the north. Several other large companies specialize in paper production for domestic markets. (Data provided by companies, field notes 1992, and Association of Manufacturers of Pulp and Paper, 1991.)

Throughout Brazil and to a lesser extent in other countries where plantations have been established, there is a great deal of research in progress on such features as cloning and planting techniques, hybrids, soils, moisture, and as well on ways of overcoming the hazards of monocultures, the impacts on climates and water-tables, and even the impacts on human communities. In this respect, as in respect of creating the raw material rather than taking it from nature, the Brazilian industry is not a repeat of the Canadian industry.

Land already in production sells pulp at prices well below those of competitors in Scandinavia and Canada. Growth limits will be determined by available land, which is why companies are now moving into the southern and even northern Amazon; and any future ecological impacts connected to monoculture plantations. These are less stringent limitations than many faced by northern producers, which is why observers believe that Brazil will be the powerhouse of the industry within a short time. One analyst, a director of a major Spanish firm, has argued that the entire world's pulp requirements could be met by trees grown on just 3 percent of Brazil's land (Wilson, 1993).

Chilean Plantations

Chile is also developing a strong pulp and paper industry, and is cutting and displacing its temperate coniferous forests in order to plant pines and eucalypts. It has less available land than Brazil, and land policies are hotly contested by native peoples and a strong ecology movement, but its openness to foreign investment, its emphasis on industrial development at all costs, and the atmosphere of repression for critics contribute to increasing plantation investments. No fewer than 15 mega-projects in the forestry sector were planned for the period 1992 to 1996, including three major pulp and paper mega-projects with joint ownership between combinations of the four or other major Chilean investment groups and, respectively, by Licanten of Switzerland, Stora of Sweden, and Daio and C. Itoh of Japan (special printout provided by Chile, Instituto Forestal [INFOR-CORFU], 1993).

Four holding companies control about 40 percent of plantations, and produce about 63 percent of forest products (Contreras, 1989). The major companies are Compania Manufacturera de Papeles y Cartones (CMPC) and Celulosa Arauco y Constitucion (Arauco). The first is controlled by Chilean capital, the second is jointly owned by New Zealand's Colt Holt Harvey and Chilean capital. CMPC, with several mills and plantations, has a total of 337,000 hectares, about two-thirds in radiata pine; and produces some 540,000 tons of pulp and paper annually. Arauco, originally established by the Chilean state in 1972, was transferred to the private sector under the Pinochet regime. Later, Carter obtained nearly 60 percent of shares in COPEC, an industrial conglomerate which in turn has majority holdings in Arauco. Financing of new mill capacities in 1989 was provided by the International Finance Corporation supplemented by a loan from the West Germany development agency DEG (Paner, August 22, 1989). The company now has three very large mills with a combined capacity of 805,000 tons per year. It rests on a total of 437,000 hectares of forest land, three quarters of which is planted in radiata pine.

Indonesian Beginnings

Although Indonesia has less genuinely marginal land already deforested and available for plantations than Brazil, it has over 40 pulp or pulp and paper mills. Some, perhaps even a majority of these, produce wood-free papers from rice straw, bagasse, bamboo, and wastepaper sources. The largest companies, however, cut tropical forests in order to plant acacia and other species for new and very large pulp mill installations. The Sinar Mas group, owned by a wealthy Indonesian family, has six pulp and paper companies including majority shares in one of the largest complexes, Indah Kiat. In 1988 Indah Kiat's vice president told Pulp and Paper International that "we are looking for forest which can be clear-cut and replaced with eucalyptus and acacia" (Soetikno and Sutton, 1988: 41). Its first mill on Sumatra rested on 65,000 hectares of tropical forest, since expanded to provide for pulp production at capacity of 790,000 tons plus 254,000 tons of writing and printing paper. A third mill in the same region draws on a 300,000 hectare concession. Pulp and Paper International reported in October 1994, that the company anticipated continued substitution of tropical hardwood with plantation hardwoods. Indeed the company, along with other companies, is rushing to beat the Rio Conference targeted date of 2000 for closing tropical rainforests to logging.

Ironically the Indonesian industry to this date is not profitable for the Indonesian economy, though it generates substantial wealth for individual groups of investors. Wahana Lingkungan Kidup Indonesia (Indonesian Forum for the Environment) argues that only 62 percent or 90.2 million hectares of the forest remain undamaged; fewer than 40 million hectares have the potential for sustainable timber production, and contributions to domestic income between 1983 and 1989 accounted for only 1 percent of Indonesia's total GDP at both current and constant 1983 market prices (1992a and 1992b). Despite these problems, the Indonesian government is determined to establish a strong plantation/pulp mill industry modelled on the Brazilian prototype, to both meet domestic demand and bring in export dollars.

Malaysia and Thailand at the Margins

Malaysia and Thailand are still on the margins of the new plantation economy. In both countries the natural forests are almost gone, and plantations will cover the land if investment funds materialize. Limited logging bans in natural forests of Thailand were imposed in 1985 and total bans were imposed by 1989, though illegal logging continues. Thailand's population increased from 18 million in 1953 to 52 million in 1986, and this reduces the capacity for extensive development of plantation forestry though the Thai government encourages investment in the sector. Community forestry (the Taungya system or variations on it) is regaining strength in parts of the country (Amyot, 1988; Sahanulu and Hoamuangkaew, 1986).

Malaysia was the leading exporter of tropical hardwood in the late 1980s, even following the ban on sawlog exports in 1985, and on all logs from peninsular Malaysia in 1989. Exports are now entirely from Sabah and Sarawak, where the harvesting rate in the mid 1980s was estimated to be four times the regeneration rate (World Wood, 1987: 45-6), and later, observers argue that such estimates were too low (Sahabat Alan Malaysia, 1990; Sesser, 1991). The objective of the Malaysian government and dominant companies is to log out the natural forest in order to plant cash crops. One mill now operates in Sabah, wholly-owned by the government, drawing on a 271,000 hectare concession area. Plans for new pulpmills and a newsprint mill are under way. The perception of natural forests is reflected in the plantation pilot scheme title: "The Compensatory Forest Plantation Project." In the government's view, companies that deforest must be enabled to plant fast-growing commercial species. The Director of the Forestry Economics Unit in Malaysia's Forestry Department decries the failure of critics to understand both "the wisdom" and "the urgency" of the policy (Sirin, 1989: 261).

Boreal Forests: Major Pulpwood Source

Boreal forests are the other major new pulp source for Asian paper companies. These include six mills already constructed or planned for northern Alberta and a Daishowa mill on the Peace River in north-eastern British Columbia. As well, the boreal regions of Asian and European former U.S.S.R. territories are being explored for potential mill sites (Cardellichio, Binkley, and Zausaev, 1990; Backman and Waggener, 1991; Tak, 1994). At the moment, they provide increasing quantities of logs to Asian markets. Boreal forests were not cut earlier by the traditional northern forest industry because they were far from markets, and the wood is generally not of high quality. But with changes in pulping technology the aspen wood is acceptable in quality, and richer lodes are already gone. Among advantages are the incentives and subsidies provided by Canadian governments, and the desperate need for investment in former U.S.S.R. boreal forest regions. The fibre comes in at very low cost. Boreal forests are good bets for the short-run, but the long-run lies with the eucalypts because they grow in 7, not 70, years.

Restructuring of European Companies

Japan is not the only country to exploit new wood sources. The entire northern industry has undergone restructuring since the mid-1970s and subsidiaries or joint ventures have been created in Spain, Portugal, Brazil, Chile and other southern climates as well as south-eastern regions of the United States. Restructuring of Scandinavian forestry was started through joint planning sessions between governments and companies in the mid-1970s. In Sweden and Finland it has taken the form of phase-outs for small, less efficient mills, retraining of the rural labour force for skilled work in more technologically advanced industries, upgrading of large mills for more value-added paper production, and relocation of pulp production to Iberian mills and paper production to mills taken over by Swedish and Finnish firms in France and Germany. The Nordic countries are taking full advantage of new rules for business in the restructured European Community. They have attempted, as well, to relocate industries in rural regions formerly dependent on forestry. Reforestation has been fundamental policy since the turn of the century, and even more intensive efforts have been undertaken to increase the growing stock well above harvesting rates (Ramoulin, 1986, 1988; Noble, 1986; Remrod, 1989; for critical perspective on these developments, J. Hansing and S. Wibe, 1993). The vital component of their approaches was concern with declining rates of rural employment and determination to re-establish employment opportunities in rural regions.

American companies have moved some of their northwest coast operations to pine and hardwood plantations in south-eastern states, and some have invested in companies in the southern hemisphere. The largest companies in the world are still American-owned, and these procure fibre on every continent. The American domestic market is also the largest market, and most American produce stays in that market.

Restructuring in British Columbia

Mill capacity exceeds the sustainable production capacity of the land in British Columbia, as reported in several reputable studies in the mid-1980s (e.g., Association of Professional Foresters, 1984; Woodbridge, Reed, 1984; Sterling Wood, 1984; Nilsson, 1985). The current government, informed now by Round Tables and CORE and the scientific panel, is moving in the direction of reducing both current cuts and future harvesting area. It is also providing incentives for companies able to produce higher value-added products that utilize wood supplies more efficiently.

A major challenge is to regain employment in regions that have been heavily dependent on the traditional forest industry. During the 1970s and 1980s computer-controlled or assisted production units were installed in pulpmills, sawmills, and woodlands, just as in other industries. Because of the expanded capacity, the impact on employment took the form of a relatively low rate of increase compared to high rates of increase in volume of production between about 1960 and 1979 (Marchak, 1983: chapter 6). Capacity continued to increase after the depression of 1980-82, but employment declined absolutely from that time to the present. Between 1979 and 1991 there was a net decrease of over 16,000 workers in British Columbia forestry, with the greatest job reductions in lumber and allied wood products. The same patterns occurred across Canada, and they occurred in every branch of the industry (Statistics Canada, annual statistics).

Local and global issues intertwine now in British Columbia. Resource depletion, declining employment and shrinking communities are conditions that would have existed if no competing industries had emerged in the southern hemisphere and boreal forests, but policies to deal with such conditions must now take the global context into account. This juncture might be seen as a time of crisis, or as an opportunity to rethink the social objectives of forest and land policies, and reconsider the relationship between the human species and the forest. It may at last be a time for escaping the staples trap.

The global context from the perspective of British Columbia industry might be understood in these terms: (1) Japan continues to demand construction wood and temperate forests provide an essential material. This would provide higher returns if producers were to market more than logs; (2) there is a growing demand for newsprint, which still requires long-fibred pulp from conifers, in South-East Asian countries; (3) there is increasing competition in pulp markets from non-traditional suppliers, especially Brazilian, and pulp can be produced by these suppliers at lower market price; (4) more competition from South-East Asian suppliers of pulp may be anticipated for the future, also at lower market price than from traditional northern suppliers; eventually these sources may supply the Asian market for pulp and most papers; (5) European paper producers are seeking lower-cost pulp sources in non-traditional regions.

This list indicates possible directions for the British Columbia forest industry. These include restructuring to produce and market more solid wood products for Asian markets, where there remain adequate supplies of good sawlogs; a deconstruction of the pulp industry in anticipation of declining market demand as Latin American companies gain market shares and Asian plantations become established. Newsprint still has strong demand in Asian markets and probably companies should shift more to them, away from American markets. Survival is more likely in niche markets for coniferous wood and pulp fibres than in markets easily and more cheaply served by southern hardwoods.

Flexible Labour Supplies

To restructure around a new product mix, take on the risks of moving into new and competitive markets, phase out some of the staples products and less competitive companies, the industry argues that it must restructure the labour force.

The current move in all industries toward "flexing" the labour force is by way of restructuring to become or remain competitive with emerging giants in the southern developing countries, in the non-union southern American states, and in the greenfield mills of Alberta. In all of these either unions are absent or agreements have not tied employers to inflexible trade definitions of jobs. But the objective of employers is not only to gain greater flexibility, it is also to produce new products with smaller numbers of regularly employed workers. Flexibility is gained by reducing the importance of credentials for some range of relatively safe and non-technical tasks; employing more of the total work-force on part-time, temporary, relief, or contract arrangements; having an "open" shop; and deploying workers to new sites or stations and new tasks from time to time.

MacMillan Bloedel becomes the test case for whether large companies can produce higher value products with the same or smaller labour force and reduced fibre supplies. Hayter and Holmes (1993, 1994) inform us that MB reduced its labour force at Powell River by 32 percent overall, and its regular hourly force by 50 percent between 1980 and 1992. It achieved more flexibility in labour policies. But its output also declined: between 1980 and 1992, its lumber production declined by 58 percent. Newsprint output was already 66 percent below capacity and fibre costs were no longer competitive for United States markets in 1980. Though new machinery was introduced, one could not deem this a successful venture in restructuring.

I assume MacMillan Bloedel chose Port Alberni for the building of a new coated paper mill because timber supplies were no longer sufficient for mass-production pulpmills. The reasoning may be: either build a higher-value-added mill or begin phasing out the site. This dilemma is built in to the whole indus-try in British Columbia. There is no longer a great deal of room to manoeuvre for either companies or labour. Flexible labour supplies are essential for com-pany survival but, clearly, reduce labour's bargaining power and the power of trade unions, and the ultimate labour force would be smaller. Companies, or at least investors, can move capital out to the new plantation territories if they so choose, and towns like Port Alberni may not survive this transition period. Given the global context and receding resources one supposes that compro-mises are unavoidable.

Plantations

There will be a continuing reduction in resource supplies from natural for-ests over the next decade and continuing non-industrial competition for good forest land. Should the province invest in plantations? Some land, especially in northern British Columbia, has no other economic value and some would argue that this section of the boreal forest has less aesthetic or spiritual value than that bordering the Athabaska if only because the Peace River dam al-ready destroyed much of the ecology and native peoples are already dispersed. The land is not useful for agriculture, and it is too far from high population areas for other industries. Its natural alder, aspen and other hardwoods are already being harvested for pulp and chopsticks. Possibly plantations could be banked against the future, but even at the most optimistic estimates of 30 years for these trees to reach commercial maturity (some foresters say 70 years is more realistic), investments in northern plantations do not look favourable compared to investments in southern plantations. If micro-climate changes or machine impaction effects have not caused soil deterioration after current logging of the natural stands, replanting could take place. But what company would tie up its capital for 30 years, possibly longer, for a fibre that may not be marketable in the technological and global context of the 21st century, especially when proven alternatives exist and a return on investment in southern plantation regions promises to be fairly quick? Plantations, however, might be the best social alternative and they could provide a portion of the domestic market. Seen from that perspective, the provincial government might be well advised to encourage plantations in northern British Columbia and provide subsidies and other incentives for that outcome.

New Industries out of Old ones

Forestry has spawned many service industries and associated technology manufacturing industries. The dominant manufacturers of pulp mill machinery are Finnish, German, British and more recently Japanese. Canadians are the innovators in logging, with self-loading barges, and various harvesting machines and trucks; also in sawmill operations with improved saws and quality control computer technologies. Few of the innovators maintained production in Canada. Instead, they moved to take advantage of capital and larger markets in the United States. But service industries in the 1980s and 1990s do not need to have physical locations in the United States in order to develop global networks, and Canadian engineering and consulting firms have become international players over the past decade. Sandwell and H.A. Simons are particularly successful, specializing in feasibility studies and pulp mill construction throughout the developing countries. Like Jaakko Poyry of Finland (which conducted the feasibility studies in Alberta), Sandwell has taken the skills and knowledge of the British Columbia forest and pulp mill industry, combined them with engineering and other specialized knowledge, and created a service that meets with a growing demand elsewhere.

There are export markets for new construction materials that use less wood, have great strength, and are less expensive than alternative construction materials. MacMillan Bloedel developed Parallam to these specifications but it is being marketed by Truss Joist MacMillan, a joint venture with an American company that is able to provide capital and market expertise as well as an established reputation in specialized construction materials.

SUMMARY

There is a very different global context for the British Columbia forest industry than existed even a decade ago. Competition in pulp markets is emerging, and new capacities in paper production elsewhere are also becoming established. British Columbia still has an edge in production of coniferous wood products, but a dwindling resource base. The issue is how should the industry restructure to use remaining resources to the best advantage; alternatively, should the province turn its back on the forest industry and reallocate the land to other uses. Restructuring of industries toward higher-value-added production is possible, but if the social objective in connection with granting resource harvesting rights is to provide employment and sustain communities, restructuring that reduces regular employment is not helpful. Restructuring the forest industry together with developing alternative industries for rural regions might be possible, but innovative spin-off industries so far have not located anywhere in British Columbia, let alone rural towns. The province may be advised to create incentives for locating in regions that need employment.

While the future of the forest industry is not likely to be as long or as affluent as its past, we have reason for optimism that at last this region might free itself of dependence on staples. Innis provided an apt description for British Columbia over the last century; respect him though we might, let us hope he is of no use for explaining the next century.

ENDNOTES

[1]   This chapter is based on material published in *Globalization of the Forest Industry*, McGill-Queen's Press, 1995.

[2]   Dr. Clark Binkley has made me more cognisant of this.

REFERENCES

Amyot, J. (1988). Forest land for people. A forestry village project in Northeast Thailand, *Unasvlva*, 159(40), 30-41.

Associacao Nacional dos Fabricantes de Papel e Celulose (1992). *Relatorio Estatistico 1991*. San Paulo.

Association of B.C. Professional Foresters (1984). Not satisfactorily restocked (NSR) lands in British Columbia. An analysis of the current situation. Mimeo, Vancouver, B.C.

Backman, C.A. and Waggener, T.S.R. (1991). Soviet timber resources and utilization: An interpretation of the 1988 National Inventory, Working Paper 35, Seattle: CINTROFOR, University of Washington in cooperation with Vancouver: FEPA Research Unit, University of British Columbia.

Cardellichio, P.A., Binkley, C.S., and Zausaev, V.K. (1990). Environmental, institutional, and economic factors limit the possibilities. *Journal of Forestry*, 88(6), 12-17, 36.

Chile, INFOR-CORFO (1992). Unpublished special printout on planned mega-projects in the forestry sector, 1992-1996, showing investment source, sector, regional breakdowns.

Contreras, A. (1987). Transnational corporations in the forest-based sector of developing countries. *Unasvlva*, 39(3-4), 38-52.

Food and Agriculture Organization (1985). Committee on Forest Development in the Tropics. Tropical Forestry Action Plan. Rome: FAO.

Hansing, J., and Wibe, S. (1993). Rationing the supply of timber: The Swedish experience. In P.N. Nemetz (ed.), *Emerging issues in forest police* (pp. 157-170). Vancouver: UBC Press.

Hayter, R., and Holmes, J. (1993). *Booms and busts in the Canadian paper industry: The case of the Powell River paper mill*. Discussion paper 27, Simon Fraser University (mimeo).

Hayter, R., and Holmes, J. (1994). *Recession and restructuring at Powell River, 1980-94: Employment and employment relations in transition*. Discussion paper 28, Simon Fraser University (mimeo).

Innis, H. (1954). *The fur trade in Canada (1930)*. Toronto: University of Toronto Press.

Japan Paper Association (1994). *Pulp and paper statistics*. Tokyo: Association.

Marchak, M.P. (1983). *Green gold: The forest industry in British Columbia*. Vancouver: UBC Press.

Marchak, M.P. (1992). Global Markets in Forest Products: Sociological impacts on Kyoto Prefecture and British Columbia Interior forest regions. In P.N. Nemetz (ed.), *Emerging issues in forest police* (pp. 339-369). Vancouver: UBC Press.

McInnis, J. (1994). *Japanese investment in Alberta's taiga forests*. Mimeo notes, 25 Feb 1994.

Nilsson, S. (1985). *An analysis of the British Columbia forest sector around the year 2000*. Vancouver: Forest Economics and Policy Analysis Project, University of British Columbia.

Noble, K. (1986). *Lessons from our neighbors of the north*. Globe and Mail Report on Business Magazine, November, 1986: 50-61.

Pratt, L., and Urquhart, I. (1993). *The last great forest: Political economy and environment in Alberta*. Paper presented to annual meetings of the Canadian Political Science Association, Ottawa, June 6-8, 1993.

Pratt, L., and Urquhart, I. (1994). *The last great forest: Japanese Multinationals & Alberta's northern forests*. Edmonton: NewWest Press.

Pulp and Paper International. Monthly, various editions as cited in the text.

Raumolin, J. (1986). Recent trends in the development of the forest sector in Finland and Eastern Canada. *Zeitschrift der Geselchaft fur Kanuda-Studien*, 11: 89-114.

Raumolin, J. (1988). *Restructuring and internationalization of the forest, mining and related engineering industries in Finland*. Etla (Helsinkl) unpublished paper No. 267.

Remrod, J. (1989). Sweden's forests, a growing resource. *World Wood*, 31(6): S4-S5

Richardson, M., Sherman, J., and Gismondi, M. (1993). *Winning back the words: Confronting experts in an environmental public hearing*. Toronto: Garamond Press.

Sahabat Alas Malaysia (1990). *The battle for Sarawak's forests*. Penang, Malaysia: SAM.

Sahunalu, P., and Hoamuangkaew, W. (1986). Agroforestry in Thailand: Present condition and problems. In Monsoon Asia Agroforestry Joint Research Team (MAART), *Comparative studies on the utilization and conservation of the natural environment by agroforestry systems. Report on Joint Research, Amona Indonesia, Thailand, and Japan* (pp. 66-92). Kyoto: Kyoto University.

Schreuder, G.F., and Anderson, E.T. (1988). International wood chip trade: Past developments and future trends, with emphasis on Japan. In G.F. Schreuder (ed.), *World trade in forest products* (pp. 162-84). Seattle: University of Washington Press.

Sesser, S. (1991). A reporter at large. Logging the rain forest. *The New Yorker*, May 27: 42-67.

Sirin, L.M. (1990). Malaysia. In Asian Productivity Organization (ed.), *Forestry resources management* (pp. 247-268). Tokyo: APO.

Soetikno, A., and Sutton, P. (1988). Time to talk in terms of millions. *Pulp and Paper International*, 30(1): 40-44.

Sterling Wood Group Inc. (1984). *Status of the British Columbia coast forest industry*. Victoria: Queen's Printer, Sept. 1984.

Tak, K. (1994). *Foreign direct investment in forest resource development in the Russian Far East and its implications for Japanese and Korean log markets*. Ph.D. dissertation, Faculty of Forestry, University of British Columbia, October, 1994.

Wahana Lingkungan Hidup Indonesia and Yayasan Lembaga Bantuan Hukum Indonesia (1992a). Mistaking plantations for the forest: Indonesia's pulp and paper industry Communities, and Environment. Jakarta, July, 1992.

Wahana Lingkungan Hidup Indonesia (1992b). Sustainability and economic rent in Indonesian forestry sector. Jakarta, 1992.

Westoby, J. (1989). *Introduction to world forestry*. Oxford: Basil Blackwell.

Wilson, J. (1990). Wilderness politics in B.C.—The business dominated state and the containment of environmentalism. In W. Coleman and G. Skogstad (eds.), *Policy communities and public policy in Canada: A structural approach* (pp. 141-69). Mississauga: Copp Clark Pitman.

Wilson, R.A. (Commercial and Forestry director of Celulosa de Asturias SA [CEASA], Spain) in *Papertree's Forum*, Papertree, Oct. 1993.

Woodbridge, Reed and Associates (1984). *British Columbia's forest products industry constraints to growth*. Ottawa: Ministry of State, Economic and Regional Development, June, 1984.

World Wood. October 1987: 45-6.

Plate 14   Coastal pulp mill   ➤
Plate 15 (overleaf)   Testing room   ➤

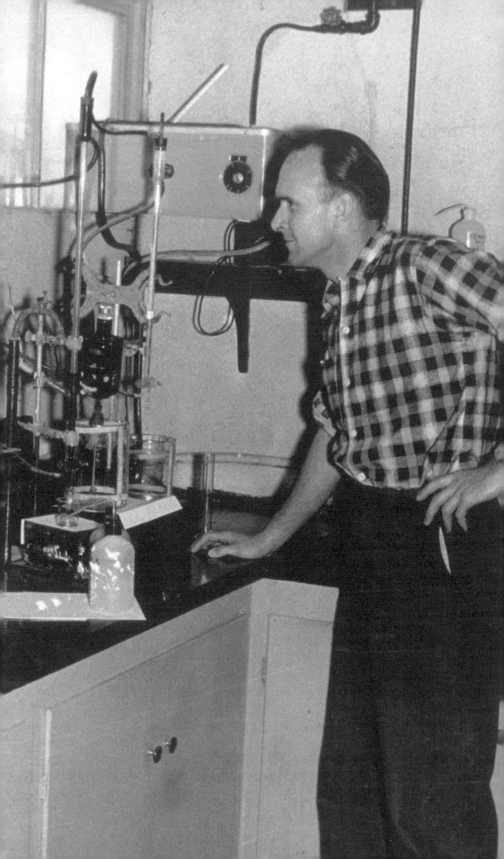

The British Columbia Forest Industry: Transition or Decline?

10

Otto Forgacs
Former Vice-President, Research & Development, MacMillan Bloedel

The forest industry is faced with an unprecedented array of challenges, stemming from globalization of production and markets, the electronic revolution, and changing social and environmental expectations. In its need to deal with the impact of globalization, computerization, and changes in markets, the position of this industry is not unique. Is there an industry today that isn't faced with such problems? Special problems for the forest industry in British Columbia arise, however, from the demographics of tree populations (distorted by the consequences of inadequate regulatory analysis in the past), the abrupt changes in the value that the public attaches to conservation and to "old growth" stands, and the way that the provincial government is attempting to deal with both. This chapter addresses how the industry is adjusting to these issues from employment and technological perspectives.

Ideally, the process of transition in British Columbia's forest industry system must be handled using the best available knowledge and much common sense. Serious errors in this process, whether they be based on political expediency, ignorance or ideology, could seriously damage both the economy of this province, its employment base, and ultimately, our ability to protect environmental values. This chapter argues that employment levels and profitability in the forest industry depend upon wood supply and the incorporation of new technology. Such an argument contrasts with prevailing environmental interpretations, which show that automation is reducing jobs in the industry. The author's argument is supported by reference to aggregate trends in the industry and by case studies of recent technological innovations

THE IMPORTANCE OF THE BRITISH COLUMBIA FOREST INDUSTRY

The forest industry, long the most significant generator of economic wealth in British Columbia, is still hugely important. As recently as 1993, its manufacturing component dwarfs all other such activities in the province (Figure 10.1).

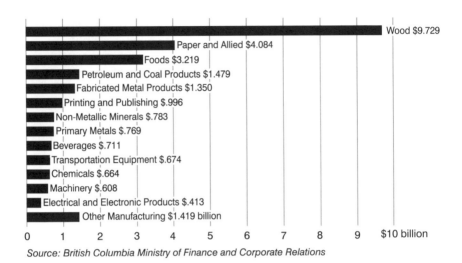

Source: British Columbia Ministry of Finance and Corporate Relations

Figure 10.1   Manufacturing shipments, 1993 (billions of dollars)

In terms of value of shipments, for example, wood and pulp and paper accounted for over half of the provincial total.

In 1993, the forest industry employed about 92,000 people directly and supported another 184,000 indirectly. One job in every six still depends on the industry. In addition, the forest industry paid 5.1 billion dollars in wages and benefits; the average annual wage before benefits was over 41,000 dollars, which is 41 percent higher than the average annual wage (29,000 dollars) in the province. In 1993, the industry and its employees also paid taxes of over 3.1 billion dollars to various levels of government.[2] The 1994 figure is not available, but it will surely be higher.

Growing Competition Puts Pressure on Profits

International competition has been an increasing factor affecting the forest industry over the last two decades, and has contributed to increasingly intense fluctuations in performance. In a cyclical economy, the highest cost producers take the most punishment and get knocked out first whenever demand and prices drop. This has put pressure on the British Columbia industry to modernize, to close obsolete facilities, and to concentrate on products for which it has an advantage. Moreover, wood production from intensively managed tropical and sub-tropical plantations will increase sharply over the coming decades. However, increased demand for wood-based building materials, as well as pulp and paper products, in much of the world, especially the developing countries on the Pacific Rim, should assure British Columbia of markets for the foreseeable future. The need for wood of the burgeoning populations in the fast growing economies of South America and South East Asia should eventually soak up most, if not all, of the new supply.[2]

Another major concern within the forest industry is the impact of electronic technologies on traditional markets. The paper industry has been monitoring the likelihood of electronic substitution since the 1970s. To date, new office equipment such as copiers and facsimile and desk top printers have increased paper usage. However, current projections indicate that we may reach a turning point around the turn of the century. While newsprint and other printing and writing grades are not going to disappear in our lifetimes, and the paper-less office is still many years away, all major investments in pulp and paper facilities have to take electronic substitution into account. Irrespective of substitution, the electronic revolution has already had a large effect on the specifications for products, and the way they are marketed and manufactured; it affects the way we work and perhaps the way we think. This is illustrated here in the sections on motivation for the new NEXGEN and Parallam technologies.

Profitability is a constant concern. Contrary to the opinion of many members of the public, the financial performance of the industry in British Columbia, when viewed over several business cycles, has been poor. For example, a Price Waterhouse survey of 52 large- and middle-sized British Columbia forest products companies and 7 manufacturers' associations reveal that the average industry returns are seen to compare unfavourably with those available by investing in Canada Savings Bonds, which are supposed to be relatively risk free (Figure 10.2). Unless the average level of profitability can be improved, the industry will have trouble attracting the financing it needs to remain competitive. Declining profits will continue to put pressure on the industry to improve productivity, which in turn puts pressure on staffing levels, work practices and other production costs that were acceptable in easier times.

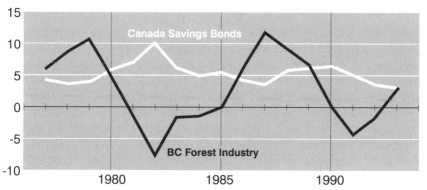

Source: Forest Industry In British Columbia - 1993, Price Waterhouse

Figure 10.2   British Columbia forest industry return on assets, 1977-1993

Job Loss—The Real Cause

A common impression is that the main cause of job loss in the forest industry is a result of the substitution of capital for labour. It is true that this factor is
ever present in industry, and is the direct result of efforts to remain competitive.
At some of the older industrial sites, such as Port Alberni, where logging curtailments have coincided with the closure of obsolete facilities, it has caused
considerable local hardship. The author wants to make it clear that he does not
want to belittle that issue, but wishes to clarify the causes. In aggregate terms,
employment trends for the industry as a whole have shown a modest decline
(Figure 10.3).

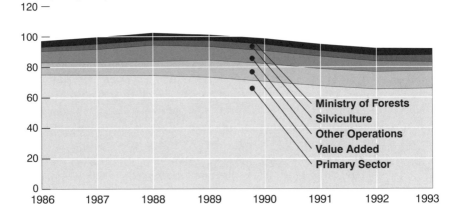

Figure 10.3   British Columbia's forest sector employment estimates
(based on adjusted Price Waterhouse data)

It might be noted that the decline in primary activities (logging, wood
processing, and pulp and paper) has been offset by an increase in employment
in the so-called "value-added" sector, otherwise known as "secondary manufacturing," which is good news. There are also increases in forest industry related government employment. Absurdly, there are more people in the British
Columbia Ministry of Forests than there are people practicing silviculture in
the province.

A look at employment and logging levels over a longer time span, between
1970 and 1993, presents an interesting and revealing perspective (Figure 10.4).
One does not have to be a statistician to identify the strong correlation between total employment and the amount of wood logged. Until the late 1980s,
these fluctuations in logging activity reflected mainly the cyclical peaks and
valleys of market demand, indicating clearly the sensitivity of employment to
logging levels.

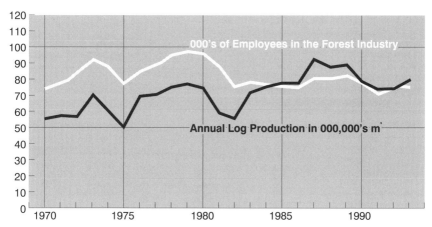

Note: Forest Industry Includes Logging, Solidwood Products and Pulp and Paper.
Source: Stats Can - Census of Manufacturers (AS) and COFI

Figure 10.4   Employment vs log production in the B.C. forest industry

Since about 1990, provincial government log allocation, rather than the market, has become the limiting factor. The level of logging is now slated to drop substantially further under the provincial government's forest policies, especially the Forest Land Reallocation Process (Protected Area Strategy and CORE), the Timber Supply Review, and the Forest Practices Code. Each of these programs has some merit, when viewed independently. But what is the impact of applying them all together and all at once? Government estimates of the impact are not clear. The government merely announced that any job losses would be offset by the Forest Renewal program, together with increases in secondary manufacturing. Is this realistic? On the other hand, a number of industry and academic estimates put the potential decline in the allowable cut at 25 percent. Price Waterhouse (1993) has calculated that 3.5 direct and indirect jobs are dependent on every 1000 cubic metres logged. Clark Binkley's (Chapter 2) results suggest that this is an underestimate, putting the number closer to 4.7. His calculations indicate that a 25 percent reduction in the level of harvest will mean a loss of 92,000 jobs, together with a 4.9 billion dollar/year reduction in the provincial GDP, and a more than proportionate drop in tax revenue.[4]

One would assume that the implications of the new programs for the provincial employment base and economy would have received thorough study before their introduction. Clearly, this does not appear to have been the case.

It should be pointed out that the job statistics have been frequently misused in attempts to indicate that automation and mechanization are the only cause of employment decline that it is really necessary to worry about. For instance, certain special interest groups have claimed that the industry lost over

27,000 jobs, as the result of mechanization and automation. This conclusion is reached by comparing the numbers for the 1979 boom year to those for the recession of 1991 (see Figure 10.4), and there are many other examples of similar errors, if not attempts to mislead. It is evident that the provincial government also has tried to associate job loss with modernization, perhaps to deflect public opinion away from the large negative impact of the proposed logging curtailments.[5]

The emphasis on job loss as the result of new technology and modernization is sometimes presented as a ruthless characteristic of large corporations.[6] Yet modernization also preserves jobs, and new technology often creates them.

TECHNOLOGY AND THE COMPETITIVENESS OF THE PRIMARY SECTOR

The following section describes some of the actions that have been taken by the primary sector to respond to markets, as well as to the need to make better use of the available fiber so as to remain competitive. The examples are drawn from the activities of MacMillan Bloedel (MB), not to single out the achievements of this company, but because the author has first hand knowledge of the projects mentioned. The author felt, during his tenure with MB, that by remaining competitive they were protecting jobs rather than destroying them. The four following cases offer support for this view.

Case I: Increased Recovery of Value from Logs

During the post World War II boom period, value recovery from logs in this province was neglected, and huge values were lost by converting prime old growth to dimension (commodity) lumber. An indication both of past practices and of current progress is the remarkable increase in value recovery that has occurred in recent years. For example, at MacMillan Bloedel, the portion of lumber fetching a premium for decorative value has grown from 13 percent in 1981 to 37 percent in 1993, despite no improvement in log quality. Other companies also report impressive progress. For example, Peter Bentley (1994) claimed that Canfor's sales of value-added commodities had almost doubled to 700 dollars between 1990 and 1993.[7] At MacMillan Bloedel, the shift to increased revenues has been achieved through better management in the forest, at the log sorts, and at the sawmills. There are further gains still to be made, as the result of advanced technologies such as, for example, better techniques to reduce insect attack on felled logs, and X-ray scanning of logs to guide value recovery. Such progress is not recognized as "value-added" by the current provincial government, but it is essential to global competitiveness of the primary sector, and increases the supply of wood available for further manufacturing.

The next two examples are included not only as illustrations of value added by the primary industry, but also because they show how the process of technological innovation works.

Case II: NEXGEN: From Undifferentiated Newsprint to Coated Speciality

High wood costs have reduced the competitiveness of coastal British Columbia mills producing commodities, including standard newsprint, which is relatively low in the hierarchy of value-added paper products. Typical of commodities, its price is volatile, depending on the world supply/demand balance.

For the papermill at Port Alberni, the move to value-added is a survival strategy. There are signs that the daily newspapers may be at the threshold of a long-term structural decline. A recent survey showed that the percentage of adults (18 and over) in the U.S. who read daily newspapers has dropped from 78 percent in 1970 to about 62 percent in 1993.[8] Young people tend to get their information from television and personal computers, and will never be the heavy consumers of newspapers characteristic of the author's generation. Consequently advertising revenue, particularly the component aimed at young people, is drifting towards electronic media and special interest magazines. This does not mean that newsprint manufacture will necessarily be a bad business for everyone, but the pressure to remain among the world's low cost producers will be fierce. Low cost wood, easy access to recycled fiber, competitively priced electrical energy, and modern, high speed equipment are the ingredients for a mill to be low on the cost curve.

A product line which has a brighter future than standard newsprint is lightweight coated paper (LWC), the clay coated product that is the preferred sheet for magazines and catalogues. This sector of the printing and publishing industry remains strong, and is projected to grow at a rate somewhat in excess of the rate of growth of the GNP. It is unlikely that special interest magazines, with their ability to target advertising, will be adversely affected by electronic alternatives within the foreseeable future. These papers should continue to grow at a rate in excess of that of the GNP.

About 10 years ago, a major study was undertaken by MacMillan Bloedel to convert one of the Alberni papermachines to LWC. The good news was that the softwood fiber available to the mill was very well suited to the production of this grade, and that there was strong demand for the product on the U.S. west coast. The bad news was that the return on such a conversion, using the then available technology, was substantially below the cost of capital.

However, the company did not give up, and started looking for a technological solution. In 1991, the mill, working closely with MB Research and a supplier in Finland, identified new technology called "metered film coating," which had been developed for other paper applications, and which appeared to hold promise that it could be adapted to the needs of the Alberni LWC project.

The project, which was dubbed NEXGEN (for "next generation"), was carried out over a period of about four years, and involved installing pilot plant equipment in Burnaby which cost in excess of 9 million dollars. The result is a potentially excellent product, using equipment that can be installed at about half the cost of conventional coating machinery. Even with the new technology, about 210 million dollars of capital was required, and NEXGEN turned out to be the largest capital project undertaken by the British Columbia forest industry in several years. At the time of writing this chapter the project was on schedule, despite much publicized differences over the choice of construction unions. Production was due to start before the end of 1995. The project is a major advance in value-added technology that should do much to keep the Alberni papermill on the map, and provide some much needed stability to this hard-hit community.

Case III: Parallam: High Strength Structural Lumber from Residuals

Parallam parallel strand lumber is a high strength engineered wood product that was first used to frame the foreign pavilions at Vancouver's Expo 1986, although it did not become commercially available until 1990. In the last five years it has become a familiar product to builders across North America.

The Parallam project is the largest new product development project ever undertaken by MacMillan Bloedel. It started more than 25 years ago as the result of the extraordinary vision and creativity of its inventor, Derek Barnes.[9] The research and development phase, carried out first in MB's laboratory in Vancouver and later in its Annacis Island plant, cost about 45 million dollars, and took over 18 years to bring from the laboratory to commercial fruition.

Even 25 years ago, it was evident that British Columbia's supply of large construction timbers would not last forever. Furthermore, it was becoming increasingly evident that the high quality clear lumber from large mature trees should be sold for its aesthetic value, rather than its structural strength. The technological challenge was to develop large structural members by engineering a wood product which did not require large, old growth trees and which would fit the needs of the North American market for heavy duty structural components for low-rise housing.

It was decided to develop a premium product that would not only meet the above objectives, but have advantages over natural wood. Such a product should be strong, should not warp or split, should be available in long lengths and large cross sections, and should be able to meet precise engineering specifications for strength and stiffness. It was later found that precise quality standards were even more desirable than first realized, because of the proliferation of computer-aided design methods that are now broadly available to builders and architects, another example of the impact of the electronic revolution.

The feed stock chosen for Parallam was softwood veneer that was unacceptable for plywood or other laminated products, such as the veneer that is produced to "round up" logs on a lathe before complete sheets are produced. While this material is waste in a veneer plant, the wood from the outermost part of the log is prime material for Parallam veneer; it is converted into approximately eight foot strands, which are then aligned, coated with resin and cured in a continuous press using microwave energy.

The first fully commercial Parallam plant was started on Annacis Island in 1989, and reached full production about 18 months later. Since that time, the plant has been expanded. It directly employs about 110 people. Today, the British Columbia Parallam plant cannot find all the wood it needs in the province, and has to import veneer residuals from the U.S. Subsequently, a second plant was built to take advantage of a wood surplus in Georgia. In 1992, MB's engineered wood interests were merged with those of Trus Joist Corporation of Boise, Idaho, in a partnership called Trus-Joist MacMillan, which is now building a third plant in Virginia.

Engineered wood products have great potential for the increased manufacture of prefabricated components for houses and industrial structures. The opportunity also exists to manufacture beautiful value-added products for construction and furniture by building strength and stability into engineered core material and covering it with high quality sliced veneer from mature softwood trees or decorative hardwoods. The opportunities for value-added products from engineered and composite wood products seem unlimited. However, plants to manufacture new materials such as Parallam are research and capital intensive. Large industry and small companies can therefore play a complementary role.

Both NEXGEN and Parallam are examples of contributions to the restructuring of the industry that only a large company can make.

The (Limited) Potential of Secondary Solid Wood Manufacturing

There are about 600 companies in British Columbia engaged in the so-called secondary manufacture of wood products. The products range from prefabricated houses to smaller components, from beams to boxes, from flooring to furniture, and from siding to stairs.

This sector employs about 12,000 people and has total sales in the order of 1.5 billion dollars.[10] Most of the operations depend on British Columbia forests, with the exception of the high-end millwork and furniture companies, which process predominantly imported dense hardwoods, such as oak and maple. Many of the companies are small, with fewer than 25 employees, and about half are located in the Lower Mainland area. There is no doubt that this sector has further growth potential, and perceived barriers to its continued development, such as access to wood, and assistance in education, training, technology and marketing, should be addressed.[11]

Detailed economic data on this sector, such as its profitability and tax revenue generated, are limited. Many of the jobs, while providing much needed regional employment, are substantially lower paid than the largely unionized jobs of the primary sector. Furthermore, the future growth rate of this sector is not easy to forecast. In Washington State, which has had to deal with similar pressures, its growth rate is about 4 percent.[12] That would imply that the current level of employment could double in about 17 years. The forces driving towards economy of scale and the replacement of low-skilled labour with sophisticated, capital intensive, numerically controlled machines, manned by highly trained technicians, will become increasingly evident. This, in turn, will create pressures to form larger units. The secondary manufacturing industry is by no means immune to the competitive pressures of the global market.

One must question, therefore, the premise of the provincial government's claim that employment in the secondary sector can compensate for the economic wealth lost to the Province through an abrupt, major reduction in allowable cut. The need for a closer analysis seems vital.

CONCLUSION

The government is intent on reducing the annual cut to meet the still somewhat ill-defined goal of "sustainability." A number of programs have been imposed on the forest industry, with apparently inadequate projections of their combined impact. Unfortunately, there is a lack of knowledge about relevant rates of change. The data show clearly that harvest level, rather than modernization, is the main variable affecting industry employment. There appears to be a vital need to apply disciplined, objective analysis to the full impact of each of the new forest policies, to assess their effect separately and in combination.

The evidence suggests that the negative effects on both revenue and employment will kick in much faster than will the effects of the proposed offsets. The impact of the silvicultural initiatives on both jobs and provincial revenues needs to be modelled.

Three examples have been provided of many actions in the so-called Primary Sector to improve competitiveness in a changing market. They show that innovative technology can maintain the industry's competitive edge. The author continues to maintain that a job saved is equivalent to a job gained, and should attract equal encouragement.

It is the argument of the author that the best interest of the Province is served by recognizing the complementary roles of large, medium, and small businesses, and policy should encourage cooperation. Some industries, by their very nature, require large corporations. A new paper machine, installed, costs about 350 million dollars (or closer to 600 million dollars if a mechanical pulping line is included), a greenfield Kraft pulp mill, complete with modern environmental technology, costs over 1 billion dollars, and even a replacement recovery furnace in such a mill has an installed cost of 160 to 200 million dollars. Small companies, or even medium-sized companies, have no place in the pulp and paper industry, something that was recognized in Sweden, where the pulp and paper industry has essentially merged into only three large corporations. To its credit, the B.C. Ministry of Employment and Investment understands well the separate and complementary roles of different sized companies, as shown in Table 10.1, but the understanding does not appear to be reflected in the actions of the B.C. Minister of Forests.

It is the author's belief that the forest industry in British Columbia has the commercial and technical skills and resources to weather this period of acute change, to emerge from the current transition, perhaps a little smaller in volume, but vibrant and globally competitive, and that the current policies of acting first, and thinking through the impact later, are leading us in that direction. The author can only echo Clark Binkley's concluding plea for mutual respect and cooperation, and for the difficult "revolution of the mind" that is required. Unless goodwill, cooperation, and reason prevail, that goal will not be realized, and that is the true crisis.

Table 10.1   Industrial structure

World class value-added industry composed of a mixture of:

- *Small companies*
    - that are independent, entrepreneurial, flexible, and located in the smaller communities
- *Medium companies*
    - that are innovative and specialized
- *Large companies*
    - that utilize advanced technologies
    - have sophisticated distribution systems
    - and can undertake large projects

with the complementary support industries

- that supply equipment, technologies, distribution services, and R&D

Source: Ministry of Environment and Investment, 1994.

ENDNOTES

[1]   Price Waterhouse, "The forest industry in British Columbia", 1993.

[2]   Sutton, W.R.J. (1993). The world's need for wood, presentation to the conference on *The Globalization of Wood*, Forest Products Society, Portland, Oregon, Nov. 1, 1993.

[3]   see 1 above.

[4]   Binkley, C. (1995). *A crossroads in the path to a sustainable forest sector in British Columbia*, Harold Innis Centennial Celebration conference, Trouble in the Rainforest, 16-18 February, Vancouver, 1995 .

[5]   Others have compared 1979 with 1990 [see, for example, *Value Added Manufacturing in the Kootenays*, Ministry of Small Business, Tourism and Culture and Ministry of Employment and Investment, Oct. 1994]. Similarly, Evans and Conroy, two MLAs who are respectively chair and co-chair of the Provincial select standing committee on Forests, Energy, Mines and Petroleum Resources used data for the years 1989, 1990 and 1991 to indicate the drop.   Even Patricia Marchak falls into the trap of selective statistics by comparing the numbers for the boom year 1979 with 1986 [Marchak, M.P., *BC Studies*, no. 90, p. 21,1991] and erroneously attributing the drop primarily to automation.

[6]   For example, Andrew Petter, BC Minister of Forest, recently stated [Petter, Andrew, Kootenay Wood Forum. Creston, BC, October 23, 1994] that "since 1965, the number of jobs per thousand cubic metres of wood harvested in this province has dropped by nearly 50 percent." It would have been more accurate to say that production has doubled over this period with little change in total employment.

[7]   Peter Bentley, CEO of Canfor, during his presentation to the Forest Sector Strategy Committee in Prince George, Nov. 1994, claimed that Canfor's value-added wood product sales had almost doubled to 700 million dollars in the last three years.

[8]   See *Facts about Newspapers, National Newspaper Association of America (1994)*. George Weyerhaeuser Jr. recently referred to data showing that the use of newspapers has a strong correlation with age. [Weyerhaeuser, G., in Proc. 7th Annual B.C. Forest Industry Conference, Price Waterhouse, March 15, 1994].

[9]   In 1987, Derek Barnes and co-inventor Mark Churchland were awarded the prestigious international Marcus Wallenberg prize for this contribution to the forest industry. This prize is presented annually by the King of Sweden.

[10] Structure and Significance of the Value-Added Wood Products Industry in B.C., FRDA Report 0835-0752; 203 (1993).

[11] Result of survey conducted for the SPARK Advanced Forest Products Initiative, Science Council of British Columbia (1994).

[12] BC Ministry of Employment and Investment, who quoted the Department of Trade & Economic Development, Washington State, 1994.

Plate 16   End of the paper machine   ➤

Plate 17 (overleaf)   Taping the paper roll   ➤

The Restructuring of British Columbia's Coastal Forest Sector: Flexibility Perspectives*

11

Roger Hayter

Department of Geography, Simon Fraser University

Trevor J. Barnes

Department of Geography, University of British Columbia

* This essay first appeared in *BC Studies*, Spring, 1997

The 1980s and early 1990s represented a sea change of sorts for British Columbia's forest product sector as the industry moved from a regime of mass production or Fordist style manufacturing to a newer one emphasizing some form of economic flexibility. Within a decade and a half production technology, markets, supply conditions, employment practices and corporate organization all significantly altered as the 'Golden Age of Fordism' was nudged, and sometimes shoved, aside by the brave new world of flexible production (for the local details see Hayter, 1987; Barnes and Hayter, 1992; Barnes, Edgington, McGee, and Denike, 1992; Hayter and Barnes, 1993; Drushka, Nixon, and Travers, 1993; and for the broader picture see Harvey, 1989).

These changes occurring in British Columbia are part of a more general global economic transformation. But while the trend is global, its precise form is finely variegated, varying by nation and sub-region, by industrial sector, and by market type and segment. Equally diverse are the consequences of economic flexibility. One is the burgeoning of high-paid, highly skilled flexible jobs, for example, in the high-tech enclave of Richmond, British Columbia (Klopfer, 1995). Another, though, is the replacement of well-paid, semiskilled occupations by lower-paid, so-called numerically flexible ones in, for example, some of the non-unionized wood remanufacturing plants recently opened in the lower Fraser Valley, British Columbia (Rees, 1993). Yet another consequence is that employment of all types is lost as firms move to practice flexibility elsewhere, as when the Vancouver manufacturer of Janzen's clothing moved to the United States Sunbelt in 1996 (for a general discussion see Rauch, 1996).

Economic flexibility, then, is not uniform, but diverse and heterogeneous. It is as much found in the East Vancouver houses of homeworkers assembling swimming goggles for offshore firms (Ocran, 1995) as it is in the Burnaby offices of Ballard Power Systems as electrical engineers peer day and night at their computer monitors.

The purpose of this chapter is to map the peculiar local form of economic flexibility found in the forest sector of coastal British Columbia. We concentrate on the coastal region because the trend towards flexibility is more evident there than in the interior, a result of older and more costly coastal mills in greater need of investment as firms modernize and try to reduce costs. Specifically, we describe three main forms taken by economic flexibility in the coastal forest industry: around production technology, and especially the use of computer-assisted machinery and associated work practices; around a new type of labour market organization based upon the degree of a worker's flexibility; and around local economic strategies of single industry forest communities as they cope with a late manifestation of that age-old tension within capitalism between the fluidity of the market and the rootedness of place (Zukin, 1991).

In developing our argument two qualifiers are necessary. First, although we stress flexibility within only the forestry sector, we are aware that the shift to flexibility is associated with a broad spectrum of political, social, and cultural issues not considered here, for example, those turning on the changing role of the state (Johnson et al., 1994). Flexibility is not a limited economic shift in production, but signals far-reaching political economic effects. Second, although we do not examine recent major policy discussions about the future of British Columbia's forest economy, such as the Peel report and the CORE process, we think the chapter has implications for them. Policy imperatives such as a stronger commitment to environmental and community sustainability, non-wood values, more diversified tenure arrangements, and the need to assign higher values to the forest resource than in the past, cannot be divorced from the changing production context. In fact, we argue in our conclusion that to meet these kinds of commitments it is necessary for the wood products sector to adopt a particular form of production flexibility that the economists Marshall and Tucker (1992) call high-performance as opposed to market-based. The transition from Fordism to flexible production cannot be taken for granted, therefore, and questions related to the organization and technology of production must be addressed along side policy debates about forest renewal, ownership, and value.

The chapter is divided into two main sections. First, we review the two general systems of industrial organization found within the coastal forest sector since the end of World War II: Fordism and flexible production. Concomitant with each, we argue, are not only distinct methods of production and labour organization, but also quite different modes of community development. Second, we interpret the recent changes in British Columbia's coastal forest sector in terms of the transition between these two systems. After briefly outlining the postwar history of Fordism in the coastal forest sector of British Columbia, and its embodiment in the single industry towns that dot the coast, we provide a detailed examination of three forms taken by flexible production.

FORDISM AND FLEXIBLE PRODUCTION

It is difficult to pinpoint the exact birth of Fordism, but Henry Ford's production of the Model-T at his Rouge River complex in Dearborn, Michigan, is undoubtedly the most well-known example. That plant embodied the two defining features of the Fordist system that subsequently came to dominate many different North American manufacturing sectors after World War II: the assembly line, and Taylorist labour relations.

Assembly-line, mass production techniques were based on the realization of internal economies of scale which, in turn, were garnered because of the production of a homogeneous product ("you can have any colour of car as long as it is black"), and the use of dedicated machinery, the specifications of which never change. In addition, the large capital requirements of Fordist production cultivated a particular kind of industrial organization, one dominated by oligopolistic competition in which large firms were vertically integrated, and often in 'arms-length' contractual relationships with suppliers.

Hand-in-glove with the assembly line went Taylorism, a particular management style and labour market organization that at its most basic involved the separation at the work place of conception from execution. Managers and engineers using time and motion studies broke down work operations into simpler tasks. "Each worker would then be assigned one task to be repeated with machine-like efficiency countless times during the day" (Marshall and Tucker, 1992: 5). In this way, Taylorism attempted both to reduce the control of workers over their own labour, and to impose a disciplinary power that was applied and monitored by a managerial bureaucracy. In turn, this management style gave rise to a bifurcated, segmented labour market, at least within large firms. On the one hand, management (and research and development professionals) formed the white collar, non-unionized, primary independent segment, and were guaranteed both employment stability and relatively high wages and benefits. On the other hand, production workers formed the blue collar, often unionized, primary subordinate segment. They also enjoyed relatively good employment stability (although vulnerable to periodic business cycle fluctuations), and, in spite of limited if any formal credentials, were paid relatively high wages based upon principles of seniority and job demarcation.

During the hey day of Fordism, the long boom from World War II to the early 1970s, the development of assembly-line techniques and Taylorism were orchestrated in North America by the three economic institutional pillars of big business, organized labour, and the state. Business and organized labour forged a mutually agreeable wage-bargain in which management organized work practices on the factory floor in accordance with Taylorist principles in return for steadily increasing wages and improvements in working conditions for workers. Such a deal worked because of the improving productivity that stemmed precisely from those Taylorist practices. In this arrangement, the role of the state was to underpin this wage bargain by specific legislative sanction, and more generally through Keynesian macro-economic management policies which ensured basic levels of demand for Fordist products. The state controlled aggregate demand sometimes by monetary policies, but more often through fiscal ones including the provision of an unemployment insurance and welfare system, along with expenditures on infrastructure.

Fordism also had a distinct geography (Scott, 1988). At both national and international scales, the large corporations that were the very stuff of Fordism produced a geographical core and periphery. Professional segments undertaking control and decisions, such as the upper echelons of management, were located in metropolitan regions, while blue collar segments were more likely to be found in peripheral areas (Massey, 1984, Malecki, 1986). Inevitably, the manner in which corporations distributed functions affected the nature of local development. In the case of the blue-collar periphery, the Fordist wage bargain struck in the 'company town,' 'mill town,' or 'union town' came to define the nature of the community itself and, in many cases, was the basis for prosperity. This last point is important because being part of the periphery means a lack of control, not necessarily a lack of wealth. As Marshall and Tucker (1992: 8) write, "workers with no more than an eighth grade education and little in the way of technical skills could end up drawing paychecks that enabled them to have two cars, a vacation cottage as well as principal residence and maybe a boat for fishing and water-skiing. The system worked for everyone."

While Marshall and Tucker's comments are primarily directed toward the United States experience, they apply equally well to the many Fordist forest-based mill towns found in British Columbia. There, high wages and employment opportunities were often complemented by desirable lifestyles organized around outdoor recreation. Certainly, there were recessions when employees were laid off, but lay-offs were temporary and accomplished through well-defined seniority rules worked out between unions and management.

Not all parts of British Columbia, or of Canada, experienced the material benefits of Fordism. In some places regional development policies were applied by the state, typically as 'top-down' intercessions in the form of grants, infrastructure investments, and tax incentives (Savoie, 1986). In effect, such initiatives were attempts to induce Fordism to areas where it was not yet found.

The places that participated in the Fordist experiment, including the forest-based communities of coastal British Columbia, enjoyed a considerable degree of stability, at least during the long boom. The mill offered secure, well-paid jobs, and even those who left the community found part-time and seasonal employment in the mill when they returned. The seeming invulnerability of Fordist resource communities was epitomized by Lucas (1971) in his now classic model of the evolution of Canadian mill towns, a model published just prior to the end of the long boom. There are four phases in Lucas's model. While the first three—construction, recruitment and transition—could be hesitant and wavering, the final one, maturity, was characterized by stable population levels, work and social relations, and linkages with the outside world. In Lucas's model, development terminates at the mature stage. While possibly describing the sociology of successful mill towns under Fordism, this model did not anticipate the implications of the global economic change that began sometime during the early 1970s, and signalled both the end of Fordism and of the applicability of Lucas's relentlessly progressive typology (Grass, 1990).

That end was heralded during the 1970s and 1980s by rapidly escalating rates of technological change, market differentiation, and global competition. Increasingly the former mutually beneficial relationship among management, organized labour, and the state was in disarray, fractured by economic changes, and by a neo-conservative political ideology that distrusted state economic intervention, and emphasized the benefits of a competitive-cost environment. In this new economic environment the sources of Fordism's former stability—dedicated production technology, highly structured labour agreements enshrined in law, and dependent single-industry communities—became the sources of its rigidities. A change to more flexible production methods, labour markets and community development was necessary. The Fordist model of mass-production was painfully transformed to one of flexible production. As a system, though, flexible production was always double-edged, with the potential to offer economic improvement, but also hardship. To make this clearer, we elaborate on three main forms of economic flexibility.

Changes in production are at the centre of the transition from Fordism to flexibility. Two main forms of change are identifiable. First, in some of the old centres of Fordism the use of new computerized technology allows the manufacture of a wider range of more specialized, high-value products. Literally by pressing a button at a computer terminal production is reconfigured, and a variety of different markets are served. Production remains capital intensive and large scale, but product differentiation is greater and the products more valuable. We term this arrangement flexible mass production. Second, market dynamism and differentiation has created all kinds of production niches for smaller firms to engage in a very fine degree of specialization, sometimes using new computerized technology and sometimes using old technology. We term this arrangement flexible specialization. Under flexible specialization small specialized firms are typically closely integrated with other firms and manifest as either a spatial cluster of highly interlinked autonomous firms subcontracting with one another to produce a final good, or as a group of specialized firms congealed around a much larger producer and undertaking specialized tasks for it, often in a close relationship. In both cases, a distinct industrial district or territorial complex is created.

Both flexible mass production and flexible specialization are usually presented as desirable. Using Streeck's vocabulary, they are learning-based forms of flexible production; forms bound up with high value-added production, employing "core" full-time workers who are already highly trained, and who deepen their skills on the job. But there is a flip side: the so-called market-based flexibility that rests on producing low value-added goods, and minimizing labour costs by drawing upon unskilled or deskilled part-time or temporary workers. This is the world of the homeworker, or of non-unionized factory labourers working for minimum wages. The French economist Alain Lipietz (1996) recently argued that there is nothing predestined about the process by which a given country or region ends up with either a learning-based or market-based form of flexible production. It is a consequence of contextual factors, which in the case of Canada, argues Lipietz, is tending towards market-based flexibility. The point here is not Lipietz's specific analysis of the Canadian economy, but his recognition that flexible production has a dark underbelly.

Changes in the labour market in many ways stem from changes in production. One of the reasons Fordism failed, some have argued, is precisely because of Taylorist management methods that separated conception from execution. The remedy is to involve workers more directly in the work process, which means that they participate in continuous training and skill widening, exercise polyvalent skills within teams and systems of job rotation, are part of quality circles, and being explicitly incorporated within such activities as supervision, monitoring, and design. As a result, the role of management is redefined. A participatory management is required: fewer managers involved in supervision, a greater focus on quality, stronger communication skills, and more commitment to provide leadership in learning about new market and product possibilities ("benchmarking"). In this way, the old line between management and workers is blurred. The two are now part of a common core group for which there are returns to education and training. In the vocabulary of the flexibility literature, the core is now defined by a set of functionally flexible workers, possessing a large range of skills and applying imagination, creativity and initiative to problems as and when they emerge. This is quite different from the Fordist world of punitive disciplinary managers, and automaton workers mindlessly repeating single, sharply demarcated tasks.

But, as before, there is another, far less hopeful side to flexible labour markets. This is the world of numerically flexible workers who are part-time, temporary, or contract workers denied the wage levels and benefits of the core. Instead, their fate is to be permanently on call, to be brought in to perform often routinized tasks that those employees in the core eschew. The labour market of flexible production can be as bifurcated as that of Fordism. Yet the division has been redrawn. It is no longer management vs. workers, but core vs. periphery.

The transformation of production and employment towards principles of flexibility has reconfigured the geography of Fordism. In communities characterized by mass production and specialization, such as Canada's resource towns, the shift to flexibility has produced substantial job loss and a sustained period of instability and flux. Certainly, any sense that resource towns had reached the final stage of maturity (as predicted by Lucas) is absent; many communities simply scramble for survival. Moreover, senior levels of government are less inclined and able to buttress ailing communities. Prompted both by increasing fiscal constraints and by a growing neoconservatism that emphasizes the market and individual initiative, there has been a move away from top-down strategies to bottom-up ones. The primary agent of change is no longer the central government, but local business coalitions often in league with other collective associations including organized labour and special interest groups.

Single industry communities were often less prepared to deal with the implications of flexibility than other types of communities because of their dependence on a single activity, the historically paternalistic relationship between management and workers, and their often isolated locations (Barnes and Hayter, 1994). Their responses, as each attempts to cope in its own way, have been diverse; Lucas's prediction of every place reaching the same end of maturity could not have been more wrong. Specifically, the variability of response among single-industry communities is occurring for three reasons. First, flexible production itself is leading to increased specialization among the towns themselves. Second, the search for flexibility is often a contentious process involving local bargaining between labour and management, varying from one place to another depending upon regional history and culture (for local examples, see Barnes, 1996). Finally, the kinds of coalitions that form at particular places to direct community development vary considerably, as do their strategies and degree of success. The result, to use Sjoholt's (1987) term, is that community development has become "unruly."

TOWARDS FLEXIBILITY IN BRITISH COLUMBIA'S COASTAL FOREST COMMUNITIES

The evolution of the British Columbia postwar forest economy gives a clear example of the transition from Fordism to flexible production. From the 1940s through to the mid-to-late 1970s, Fordism ruled supreme. A few homogeneous products were mass produced for export, namely construction grade lumber, kraft pulp, and newsprint which were manufactured in large capital-intensive operations often controlled by multinational corporations. In many ways, British Columbia was geographically right for Fordism. The seemingly unlimited old growth forests of the province combined with an insatiable demand for standardized wood products, and readily available investment capital, especially from the United States, virtually guaranteed Fordism's success here.

Other Fordist industrial features included labour relations which were along Taylorist lines, with strong unions and collective bargaining firmly in place by the early 1950s. Again, in part, Fordist collective bargaining was so successful —British Columbia's unionized workers were paid on average 5 percent more than their counterparts in the rest of the country (Copithorne, 1979)—because of geographical considerations: the value of the coastal wood stock was so high. In addition, labour was segmented with production-line workers part of the primary subordinate class, and managerial staff in the primary independent one. This bifurcation in the labour market was also matched by a dual spatial division of labour. Much of the control and decision making occurred in Vancouver, while the production function was dispersed among the many single industry communities throughout the hinterland.

In all kinds of ways the local state aided and abetted this form of resource-based Fordism. The Social Credit government in power during much of this period facilitated Fordist production by forest tenure arrangement that favoured large companies, by the construction of economic overhead capital, particularly roads and dams, and by underwriting new single industry communities as in the Instant Towns Act (1965).

During this period of the long boom, community development also proceeded according to plan. Because of the extraordinary growth of the British Columbia forestry sector, communities such as Port Alberni, Chemainus and Powell River recorded steady employment and population growth, as well as increasing affluence (see respectively: Hay, 1993; Stanton, 1989; and Hayter and Holmes, 1994). Port Alberni, for example, consistently ranked among the top 10 communities across Canada in per capita income. For a period, Lucas's model seemed to be working, at least for British Columbia's coastal forest communities.

From the mid 1970s onwards, evidence began to mount that the structure of British Columbia's forest industry was vulnerable to changing global technical, market, and wood supply conditions, and that its fibre base was deteriorating. That evidence became starkly clear with the recession in the early 1980, the most severe since the 1930s, and later with trade and environmental conflicts. In response to these difficult conditions the British Columbia forest industry increasingly promoted flexibility.

Production Flexibilities

Production flexibilities in British Columbia's forest economy, as elsewhere, are of two main kinds: flexible mass production and flexible specialization. Existing large- and medium-sized corporations dominate the shift towards flexible mass production, which is distinguished from traditional Fordism not by the scale of operation, but by the computerized manufacture of a wider range of differentiated products in which higher quality is achieved by more precise specifications and/or superior performance. Even though manufactured in large volumes, the output from flexible mass production is more finely tuned to special market needs and specific customers, and products are packaged in relatively small consignments to provide protection during transportation, and to emphasize product differentiation.

Among established major forest product corporations, MacMillan Bloedel (MB) is at the forefront of introducing flexible mass production in both paper and wood processing industries. In pulp and paper, for example, MB has substantially reduced its tonnage of commodity newsprint at both its Powell River and Port Alberni mills, and at the latter discontinued market pulp (1993) and paperboard production (1982) altogether. At the same time, the Port Alberni and Powell River mills have progressively shifted towards the production of speciality papers. In 1993 almost half of Port Alberni's production was in 40 different grades of speciality papers, including extremely light weight papers and telephone directory papers. At Powell River, which during the 1950s and 1960s concentrated almost exclusively on newsprint for the United States market, by 1993 was producing 50 different grades of paper, with about 30 percent of its production in the form of 'hi-brite' papers for use in newspapers, weekend supplements, advertising flyers, and unbound catalogues. The world's largest newsprint producer in the 1950s, the mill now has just three newspaper manufacturing machines, and with the recent decision to convert one of those it will become predominantly a speciality paper mill.

The same shift towards flexible mass production is evident among MB's wood processing facilities, a shift which began with the modernization of its sawmill division at Port Alberni in 1980 (Alberni Pacific Division). Until 1980, MB's numerous sawmills and plywood mills were large volume producers of commodities, and were becoming increasingly unprofitable. Since 1980, MB has divested its plywood operations and redefined itself as a value-added and large volume manufacturer of lumber products rather than commodities. All of its surviving sawmills have been modernized, or entirely rebuilt, to manufacture a wider range of higher quality and priced products for more specific market segments, while new mills have been established to produce new products such as parallem and other engineered woods. Moreover, MB has increasingly specialized its production by location. On Vancouver Island, MB's three 'white wood' mills at Port Alberni, Chemainus, and Nanaimo complement each other in terms of log species, quality and species utilized, and products manufactured. While the recent corporate strategy in the industry has been one of individual mills competing against one another as cost centres, here the mills have formed a 'white wood' team primarily to cooperate in marketing their products, especially to Japan. Indeed, over the past 10 years MB has been committed to penetrating the difficult, quality conscious but high-priced Japanese market, and by the early 1990s MB accounted for almost one-quarter of provincial lumber exports to that country, with coastal mills exporting about half of their output to Japan.

As with other major forest product corporations, MB's new strategy of the 1980s towards flexible mass production was crisis-driven, and began with the recession of the early 1980s. But while economic crisis was the context for MB's rethinking, it should be recognized that the firm's pioneering strategies of flexible mass-production were also facilitated by substantial in-house research and development (R&D) (Forgacs, 1993; see also Hayter, 1987). Many of the speciality papers manufactured at Port Alberni and Powell River, including the hi-brite class of papers, telephone directory, and related lightweight papers, were researched and developed in MB's Vancouver (now Burnaby) laboratories, while parallem is a new wood product developed by MB following a 20 year, 50 million dollar investment.

Complementing the shift towards flexible mass production are strategies of flexible specialization. Emerging over the last decade, flexible specialization is frequently associated with uncertain markets in which small- and medium-sized firms are low-volume, niche manufacturers within production systems that are highly integrated as a result of subcontracting linkages, information sharing, and the use of common services. In the coastal British Columbia forest economy, flexible specialization is primarily represented by the emergence and strengthening of secondary value-added wood processing activities, notably remanufacturing engineered building components, millwork, and other wood produce industries (Rees, 1993). The largest of these industries is remanufacturing that utilizes lumber from the primary wood mill to make a variety of speciality products such as door and window components, interior and exterior panelling, decking, and lumber of various dimensions. Remanufacturers are typically small, specialized firms that produce small batches of products under conditions of limited timber supply as well as market uncertainty. In contrast to the large forest product corporations, the competitive advantages of small remanufacturers are based on "substantially greater production flexibility, together with a more entrepreneurial approach and a lower overhead/labour cost structure" (Woodbridge and Reed Associates, 1984: ii). This flexibility is achieved by specialization and reliance on subcontracting for particular products (of both low and high value), and particular services such as sawing, dry kilning, and planing of varying degrees of sophistication (Rees, 1993).

In contrast to flexible mass production, flexible specialization encourages geographic concentration or clustering of entrepreneurial firms to facilitate personal contact, subcontracting, market access, appropriate labour pools, and access to common services. The main concentration of remanufacturing firms, for example, is in the Vancouver metropolitan area, but a secondary concentration has emerged in the Okanagan region (Rees, 1993). Another smaller scale but more localized example of the clustering of flexibly specialized firms is provided by Chemainus, which, since the early 1980s, has attracted a number of specialized functions, including wood component and furniture manufacturers, planers, resawers, and dry kiln operations. Chemainus, in fact, is now cited as the dry kiln capital of British Columbia, and while some of its dry kiln facilities are integrated with large-scale sawmill operations, others subcontract to major sawmillers throughout Vancouver Island and occasionally in the Vancouver area (in which case lumber is transported on the Nanaimo ferry for dry kilning and then shipped back).

Clearly, the strategies of flexible mass production and flexible specialization are related. Among the population of small- and medium-sized firms that comprise the remanufacturing industry of the Vancouver area, for example, there is at least one speciality product branch plant owned and controlled by an established corporation, Canfor, while Interfor organizes the manufacture of speciality products entirely through subcontracting. There are also subcontracting linkages between mass producers and small specialized suppliers. At Chemainus, at least, a couple of the latter were established by former managers of mass producers, including Paulcan, whose owner was a former manager of MB's old Chemainus sawmill.

Employment Flexibilities

The strong shift from traditional Fordist mass production towards flexible production occurring in British Columbia's coastal forest economy is matched by a trend towards various kinds of employment flexibilities. But there is a lot of geographical variation. Within small- and medium-sized firms pursuing strategies of flexible specialization, employment of non-unionized labour is predominant, and wages, non-wage benefits, and working practices take on different forms according to market roles and managerial preferences (Rees, 1993; Rees and Hayter, forthcoming). However, the change towards flexibility among unionized labour is especially important. In British Columbia's forest economy, more employees are unionized than are not, union contracts establish yardsticks for the non-union sector, unions have the capacity and clout to represent workers effectively, and British Columbia's forest unions are part of a Canadian tradition of unions that recognizes broader social obligations than, say, does US-based business unionism. As a result, employment flexibilities depend not only on appropriate government and business actions, but also on the actions of unionized labour.

Unionized labour, however, is not homogeneous, and differences among local branches of the same union create distinct geographical outcomes, which can also be buttressed by physical advantages of particular places. An illustration is provided by the experiences of two large, export-oriented sawmills in Chemainus and Youbou, both in the Cowichan area of Vancouver Island

(Hayter, Grass, and Barnes, 1994). In 1980, both mills employed 600-700 workers in mills that were obsolete. Since then, they have introduced new technology, diversified their markets outside North America, rationalized, downsized their workforces, and sought more flexible labour practices (Table 11.1).

Table 11.1   The Chemainus and Youbou sawmills:
Employment and production 1980-1995

| Year | Chemainus | | Youbou | |
	Employment	Production (mfbm)	Employment	Production (mfbm)
1980	650	167	655	128
1981	550	135	615	113
1982	450	36	466	133
1983	0	0	413	150
1984	0	0	413	155
1985	125	69	360	154
1986	125	69	350	96
1987	130	102	350	159
1888	135	105	350	144
1989	140	101	176	140
1995	150	101	224	52

NOTES: Employment figures are year end totals.
              Mfbm refers to million foot board measure.
Source: Hayter, Grass and Barnes, 1994: 31.

At Youbou, technical change and rationalization occurred as a parallel *ad hoc* process because of strong resistance to change by the local branch of the main union, the IWA. As of 1993, flexibility concessions were limited to a 'double decking' agreement by which employees agreed to run either mill A or mill B, whereas previously the mills had been run by completely different crews. Even so, the Youbou sawmill, already a marginal operation by virtue of its inland location, has not been profitable, and its limited employment flexibility practices have perhaps reflected, and reinforced, this marginality.

In contrast, at Chemainus, the old sawmill was torn down, and replaced two years after closure with a new mill that was fully computerized and able to cut lumber precisely to a wide variety of dimensions, particularly for the Japanese market. From its start up in 1985, the new mill was fully committed to the principles of functional flexibility with its emphasis on team work, the development of multiple skills, pay-for-knowledge schemes, ongoing training, and close management-worker interaction. In addition, the apprenticeship program was reintroduced for trades occupations (but eliminated at Youbou in 1987), and profit sharing was recently introduced. With its tidewater location and access to high quality logs, the Chemainus sawmill since reopening has been consistently profitable, and operated without lay-offs even during the recession of 1991. In the terms we used earlier, the Chemainus sawmill successfully shifted from Taylorized Fordist mass production principles to flexible production, including a move towards a core group of functionally flexible workers (for more details, see Barnes and Hayter, 1992). The union accepted the comprehensive introduction of functional flexibility principles, and the Chemainus local operates according to a sub-agreement within the provincial-wide master contract. Admittedly, in negotiations management had the upper-hand in that the old mill was closed, the workers laid off, and a clause in the master contract tied the firm to seniority only until two years following closure —that is, the firm had discretion to hire whom it wanted when opening the new mill. Nevertheless, the Chemainus example demonstrates that functional flexibility is possible within a unionized environment. That said, there may be accompanying forms of internal conflict and trauma (nearly 500 workers were permanently laid off at Chemainus), and, as a model, it need not be accepted everywhere even by the same union (as Youbou demonstrates).

Let us elaborate further on the conflicts and traumas of introducing flexible employment practices for, as noted earlier, they go to the 'dark underbelly' of flexibility. MB's Powell River paper mill provides a particularly good case study (Hayter and Holmes, 1994; Hayter, forthcoming).

In December 1973, the mill employed 2600 people with all the hourly staff belonging to one of two union locals. Following more than two decades of intermittent recession and restructuring, employment at the mill by early 1994 was only 1275 (including 235 on relief) (Figure 11.1). In spite of the 60 percent drop in employment, production capacity remains at 80 percent of its former level, and is of a higher dollar value.

To a significant degree, this job loss is explained by the effects of technological change and rationalization, including the introduction in 1990 of new, flexible work arrangements. Specifically, the shift to labour flexibility at Powell River involved modifying job demarcation lines among the trades, and between trades and production line workers, speeding up maintenance work completion times, flattening out the organizational structure of the mill to increase managerial efficiency, facilitating labour-management interaction notably by assigning more responsibility to labour, making entry requirements more rigorous to ensure that new recruits are able and willing to be functionally flexible, and contracting out. Job rotation and further reductions in job demarcation continue to be discussed.

Employment Levels

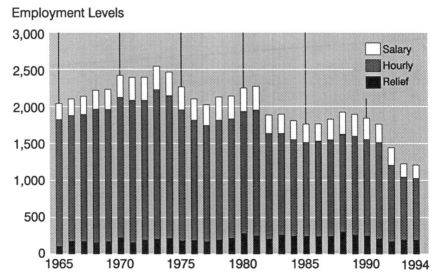

Note: The data pertain to the situation on December 31st except in 1994 when the data was for January 31st.
Source: Powell River Mill Records

Figure 11.1   Powell River Paper Mill: Employment levels, 1965-1994

There are a number of important issues here. First, in a union work place such as Powell River, flexibility, regardless of the model driving it, is bound to be contentious, not only because it demands previously 'hard fought' concessions from workers, but because it strikes at two central principles of modern unions, job demarcation and seniority. For unions, such principles serve to eliminate wage competition among workers, constrain managerial autonomy, provide security and discipline among workers, and prevent arbitrary job intensification. Contracting out is also a potential threat to these values, as well as to employment standards. Even forms of functional flexibility that stress enskilling and job satisfaction raise legitimate concerns, and any shift to flexibility demands trade-offs that are difficult for unions to make.

Second, flexibility concessions in unionized workplaces require formal negotiations between management and unions, whether these concern individual jobs and workers or are made across the board. At Powell River, for example, workplace flexibility has so far been negotiated in two agreements in 1991 and 1992 involving specific *quid pro quos*. Following a wild cat strike in 1988 over contracting out, a court action required the union to pay MB over 4 million dollars. Instead, in 1991 the union agreed to flexibility concessions in lieu of the fine. In 1992, the union then agreed to a further set of flexibility concessions in return for an early retirement package. Since then, management has been unable to offer workers another acceptable *quid pro quo* in return for more flexibility.

Third, the negative effects of downsizing on worker morale and trust (and potentially on productivity) can be compounded by flexibility discussions, especially if protracted. In the case of Powell River, the fact that agreement to be more flexible has not provided job security for the surviving workforce became a problem shortly after the 1992 flexibility concessions when another round of lay-offs occurred and the jobs the unions thought saved disappeared. For management, job flexibility was traded for early retirement with no implications for job security. For the unions, agreement to an early retirement package and increased job flexibility was traded for job security.

Fourth, in an established Taylorized workplace, many managers and workers may not have the appropriate attitudes and skills (let alone formal qualifications) for a more flexible operating culture. At Powell River, for example, there is some evidence that new, more flexible managers are not highly regarded for their knowledge about the mill, and there are problems in implementing a flatter organizational structure. Managerial inexperience with employees and machinery is likely an obstacle in the difficult process of implementing flexible work practices.

Finally, there is the issue of training. Cultivating a core workforce of stable and well-paid workers possessing polyvalent skills requires an ongoing commitment to education and training. Yet, in an old, downsizing mill such as Powell River, 'skill formation' is problematical. The apprenticeship program was a victim of downsizing. The team concept in the wood room experienced problems because of difficulties in training everybody to the level necessary to practice job rotation, and because extensive job bumping disrupted traditional on-the-job training. The lack of articulation between the new managers and the work force also poses problems for effective interactive learning. For the work force, commitment to ongoing training and education raises the spectre of 'testing'. For senior management, training and education takes people away from their jobs, and is expensive, selective, and a potential source of discontent among those not chosen.

Training is not a magic wand that can be waved to move the mill from Fordism to flexible production. Rather, training itself involves significant costs, uncertainties, and negotiation. At the same time, there is some common ground between management and labour. Both, for example, recognize the importance of entry level qualifications, and of ensuring that new entrants have general skills (numeracy, literacy, computer use, interpersonal relations, and so on) that can be built upon as needs arise. Both management and labour also suggest that there should be less reliance on on-the-job training, and a correspondingly greater commitment to more formal training involving seminars, workshops, study sessions, and experience. In spite of different perceptions about current levels of training, it is striking that management and labour are both enthusiastic about training, or, as Streeck (1989) would term it, about the firm as a 'learning institution'. The problem is to find a mutually acceptable process leading to that common end.

In sum, although all the examples are taken from just the one mill at Powell River, they make the point that the move to employment flexibility, even of the 'good' kind, is fraught with problems. Inevitably, there are large lay-offs, and the very process of implementation has the potential to undermine significant social gains won by labour during decades of struggle. Whatever the end result, the transition will be drawn out and contested. Furthermore, it will take on a definite geography, a result of historical relationships within a place between unions and management, availability of and accessibility to resource supplies, the costs of replacing sunk capital, and specific corporate location strategy. The consequence will be a wide variation in resource community development.

Flexible Communities

Under Fordism the geographical character of resource communities was fundamentally shaped by their dependence on one or two dominant employers, and by their relative isolation from major centres of population (Randall and Ironside, 1996). Admittedly, the degrees of community dependence and remoteness were not uniform, and there were also marked variations in the employment relation (for measurements and comparisons of single industry town isolation and dependence, see Randall and Ironside, 1996). The transition from Taylorized mass production to more flexible production and labour strategies within the forest industry has, if anything, increased the variability among communities. Apart from the differential effects of employment downsizing of dominant mills, flexible mass production also brought locally distinct product market specialities and working conditions, and reinforced recently in the pulp and paper industry with the replacement of industry-wide collective bargaining (the norm for more than 40 years) by mill-by-mill bargaining.

The re-articulation occurring between mills and their associated communities in British Columbia's coastal forest economy takes on various dimensions. Most obviously, the transition to flexible production has meant that once dominant companies provide communities with less income from taxes, fewer permanent jobs, fewer spin-offs for local businesses, fewer goodwill contributions, and the virtual elimination of casual (weekend and summer) employment for high school and university students. In contrast to the situation in the 1970s when dominant mills provided high-income jobs for high school drop outs, new jobs in these mills have virtually disappeared (Behrish, 1995). For example, at Powell River the paper mill has a large pool of laid off workers with recall rights, and recent liberal early-retirement packages mean that the current workforce contains few workers close to retirement age.

As dominant mills decline and company paternalism becomes less important, forestry communities are necessarily more actively involved in planning local economic development and diversification. The results are very diverse (Barnes and Hayter, 1994). There are entrepreneurial initiatives of one kind and another, and various forms of cooperation among federal, provincial, and community public sector agencies, and between these agencies and private sector agencies and unions. In some places, nothing much has happened, in other places there has been considerable transformation (for example, compare respectively Youbou and Chemainus; Hayter, Grass, and Barnes, 1995). The effects of these trends also vary, as have the economic sectors mobilized in pursuit of local economic development: tourism, education, health, transportation, housing and secondary manufacturing. In addition, many coastal forest communities are attractive to retirees and city commuters and, despite substantial industry downsizing, places like Port Alberni and Powell River have retained population, and some such as Chemainus are even rapidly growing (from 2,069 in 1981 to 3,900 in 1991).

A number of factors explain this variation in community development. Members of each community have different abilities to become pro-active in local development. Communities such as Chemainus, Powell River, and Port Alberni have had to learn to promote themselves, identify priorities, coordinate expertise within the community, deal with other levels of government, and implement plans. In Chemainus, for example, a strong business coalition linked to the mayor and city council was able to push through, with provincial financial assistance, the 'Festival of Murals', which, according to some at least, became the basis of the town's 'renaissance' (Barnes and Hayter, 1992). In other communities, such as Youbou and Port Alberni, the coalition was weaker, and the outcome less favourable (respectively in Grass, 1990; and Hay, 1993).

A second factor turns on the vagaries of changing markets. Communities are competing for finance and new entrepreneurs, while the shift to flexibility within the forest industry increasingly invokes comparisons of productivity among mills within the region and beyond. MB, for example, debated for some time whether NEXGEN should be in Powell River or Port Alberni, and even contemplated establishing a newsprint mill in northern California.

While competitive, market-driven forces divide communities, there are forces tending towards cooperation. Within the forest industry, resource rights, the locally specialized nature of investment, the size of sunk capital investments, and the presence of unions all help to regulate competition between places so that community development is not necessarily a zero sum game. In fact, important complementarities are occurring among communities, especially in tourism (for example, the Cowichan and Pacific Rim 'trails' that tie together Chemainus, Youbou, and Port Alberni). Community development at any given place, then, depends as much upon the possibilities of cooperation as it does upon the duress of competition.

Yet another factor is the capacity of the community to provide training and education now that once-dominant mills no longer provide entry-level jobs. In Powell River, for example, there has been considerable adjustment in the high-school system and among students themselves (Behrish, 1995), while other communities have grappled with adult education and retaining for those who have been permanently laid off from the forest sector (Hay, 1993). But fundamental questions remain: education for what? What will newly trained workers do?

Geography runs through all these factors. Community development is highly sensitive to where a place is located (its situation), and the resources found there (its site). With respect to situation, the degree of geographical isolation can be key. Communities that are most isolated, such as Gold River and Port Alice on northern Vancouver Island, have fared less well in the face of restructuring compared to towns on the southern island, as well as those on the mainland close to Vancouver. For example, Chemainus benefits from its relative location between Victoria and Nanaimo and the interaction between them; Youbou, with its proximity to Victoria, is now a bedroom community for the capital; and Squamish, because of its favourable situation between Vancouver and Whistler, is able to gain from the resulting tourist traffic (Reed and Gill, forthcoming). Distance of single industry towns from the major metropoles of the province makes a difference; a variant of the old core-periphery model still applies.

Many of the important site characteristics have been mentioned already: industrial location characteristics can favour a company reinvesting in the mill rather than scrapping it—for example, the tidewater location of Chemainus; the size of existing sunk capital, and the presence of an experienced and skilled workforce can have the same effect; and historic, or scenic, or recreational characteristics of a place can encourage tourist development. In addition, especially important recently is the potential within communities for commercial and residential redevelopment, a critical condition for attracting retirees and commuters (the issue of relative location also overlaps here). In-migrants have helped stabilize the populations of communities such as Powell River and Port Alberni. While it may be anticipated that new residents will occasionally 'clash' (Blahna, 1990) with old ones, newcomers add vitality, income and ideas. These new population dynamics are varied and, of course, contribute to the diversity of the community experience.

To summarize, just as production methods and working practices underwent the jolt of change with the introduction of flexible production, so, too, did the communities in which that production and work occurred. Flexible methods and workers produce an increasingly diverse set of manufactured goods in an increasingly diverse set of single industry communities. The paradox is that while flexible production lies behind the most recent round of globalization and the integration of international markets (Harvey, 1989), it also lies behind the growing dissimilation of places such as is occurring in coastal British Columbia. Geography matters at every scale.

CONCLUSION: LEARNING AS A RESPONSE TO FLEXIBILITY

The shift to flexibility is a juggernaut that will not be stopped. There can be no return to past Golden Ages, Fordist or otherwise. What we might hope for, however, is a socially acceptable form of flexibility that is fair and mutually beneficial for management, unions, and local communities. A useful distinction here, one already made, is that between market and high performance flexibility (Marshall and Tucker, 1992; Streeck, 1989). While market flexibility is about immediate cost reduction by applying an extreme form of flexibility primarily to workers, high performance flexibility involves manufacturing high value, quality products by employing high-wage, high-skilled labour. This form of employment thrives on education and training, integrated research and development, and forms of work experience that emphasize individual initiative, creativity, and imagination.

The important regional implication is that only those places dominated by high performance flexibility will prosper. In a globally competitive world, high wages will be sustained by skill, innovation, and productivity. From the examples already cited, elements of high performance flexibility are already emerging in British Columbia's coastal forestry sector. Equally, though, there are examples of market flexibility reflected, for example, in increasing contracting out, and in the often virulent labour disputes associated with it. The problem is to devise strategies that steer the British Columbia forest sector in the high performance direction, where investments in research and development, skill and training consistently seek innovative change. This strategy will not be easy to achieve. The three main institutional actors—management, organized labour and the state—each face conflicting motivations, and it is by no means certain that they will make the right choices (see Table 11.2).

For firms, core groups of functionally flexible, well-educated, and highly skilled workers are sources of innovation, productivity, and adaptability. Such a workforce, however, has various costs. As already illustrated, there is a problem of recruitment, as well as of persuading existing workers, especially if unionized, to accept the new regime; there are difficulties of ongoing training; and, partly because of riskiness and partly because of international forces of competition, there is always the inexorable pressure to lower costs of production, most immediately by labour market flexibility. For individual firms, the exigencies of recessions and the ability to poach skilled employees as needed also provide a rationale for under-investment in skill (Streeck, 1989).

Table 11.2   High performance strategies: Conflicting impulses

	Advantages	*Problems*
Firms	Innovation, productivity, adaptability	Cost of programs, loss of apprentices, uncertainty (e.g., raiding of skilled workers)
Unions	Job satisfaction, adaptability, security, high wages, public support	Trainability of members, testing fear of multi-tasking, competition among workers
Government	Acceptability of supply side policies; logic of promoting higher, more equitable incomes	Conflicting impulses hard to reconcile; potential loss of union support

NOTE: High Performance Strategy implies firms organized as innovative, learning
systems emphasizing in-house R&D, continuous training, functional flexibility
and flat decision making structures. Also implies consistent government policies
which support science and technology, education and training.

For unions, training and skill potentially provide for greater job satisfaction, adaptability to change, employment security, and high wages. Yet there is no doubt that unions face all kinds of problems regarding their commitment to the training for functional flexibility. The selective nature of training policies, the trainability of existing members, and multi-skilling (a euphemism for multi-tasking?) are all problematic, while work systems based on skill and qualifications potentially undermine the principles of seniority and job demarcation that unions have deemed essential to take competition out of the work place. Yet at a time of falling membership and of declining public support for traditional union concerns over distributional issues such as wages and benefits, training provides a potentially significant and socially desirable 'supply side' role for unions. Streeck (1992: 264-66) suggests that unions should demand high wages; encourage relatively flat wage structures with limited job demarcation and pay scales that reward knowledge rather than activities performed; demand obligatory, standardized work place training curricula, and establish proper enforcement mechanisms to ensure the curricula are followed and that training is not absorbed in production; fight for job security; pursue anti-Taylorist policies; and negotiate for training and retraining plans that meet broadly based (firm, industry, and community) needs, as well as specific requirements in programs that are ongoing rather than responses to particular emergencies.

Finally, there is the state, municipal and provincial. The provincial NDP government has been keen to promote the high performance variant of flexibility by increasing funding for skill training and job upgrading. Similarly, local resource communities have taken various kinds of initiatives to enrich locally available educational and training opportunities, not only temporarily to deal with specific problems of adjustment following large scale lay-offs, but also permanently within the school and post-secondary system. Such initiatives are to be applauded and herald a potentially significant change in attitude within resource communities, especially when taken together with signs of parallel shifts within firms and unions. That said, making single industry communities into places of learning so that they can also be places of production is difficult. First, there is the cost, especially given considerations about the deficit and increased public expenditures. Second, there is the risk for an NDP government of upsetting its traditional union allies. Third, any investment in this context is risky, it cannot be clear exactly what form this training should take or the nature of the jobs (if any) when training is completed.

Nevertheless, in principle the government, unions, and business should share a strong common interest in education, skill formation, and training, and this mutual interest should be the basis for a partnership that is in society's collective interests. As Streeck (1989) advises, firms need to shift from organizations of production to organizations of learning. Similarly, communities need to become places of learning in which families, businesses, the local public sector, and various voluntary alliances, as well as schools and formal institutions of higher education, form reinforcing networks that encourage a culture of learning. Old attitudes that education is solely a function of schools or, worse, that schools simply provide expensive day-care, should be discarded. Undoubtedly the ability and willingness of communities (and firms) to transform themselves into centres of learning will vary.

Flexibility is becoming a new way of life; whether it is good or bad is yet to be seen. There are no certainties, but, whatever the outcome, it will be largely a made in British Columbia solution, a consequence of a particular constellation of institutions and policies within the province, and of a set of compelling geographical relationships.

ACKNOWLEDGEMENTS

We would like to thank Cole Harris for his encouragement and astute comments, and we are very grateful for the financial support of SSHRC, especially for research grants 410-87-0990 and 410-91-1763.

REFERENCES

Barnes, T.J. (1996). External shocks: Regional implications of an open regional economy. In J.N.H. Britton (ed.), *Canada and the global economy: The geography of structural and technological change* (pp. 48-68). Montreal and Kingston: McGill-Queen's Press.

Barnes, T., Edgington, D., McGee, T., and Denike, K. (1992). Vancouver, the province and the Pacific Rim. In G. Wynn and T. Oke (eds.), *Vancouver and its region* (pp. 181-200). Vancouver: University of British Columbia Press.

Barnes, T.J., and Hayter, R. (1992). 'The little town that did': Flexible accumulation and community response in Chemainus, British Columbia. *Regional Studies, 26*, 647-63.

Barnes, T.J., and Hayter, R. (1993). British Columbia's private sector in recession 1981-86: employment flexibility without trade diversification? *BC Studies, 98*, 20-42.

Barnes, T.J., and Hayter, R. (1994). Economic restructuring local development and resource towns: Forest communities in coastal British Columbia. *The Canadian Journal of Regional Science, 17*, 289-310.

Barnes, T.J., Hayter, R., and Grass, E.M. (1990). Corporate restructuring and employment change: A case study of MacMillan Bloedel. In M. de Smidt and E. Wever (eds.), *The geography of enterprise* (pp. 145-65). London: Routledge.

Behrish, T. (1995). *Preparing for work: A case study of secondary school students in Powell River, BC*. Unpublished MA thesis, Department of Geography, Simon Fraser University.

Blahna, D.J. (1990). Social bases for resource conflicts in areas of reverse migration. In R.G. Lee, D.R. Field, and W.R. Burch (eds.), *Community and forestry* (pp. 159-78). Boulder, CO: Westview Press.

Copithorne, L. (1979). Natural resources and regional disparities: A skeptical view. *Canadian Public Policy, 2*, 181-94.

Drushka, K., Nixon, R., and Travers, R. (1993). *Touch wood: BC forests at the crossroads*. Madeira Park: Harbour Publishing.

Forgacs, O. (1993). MacMillan Bloedel uses product development to adapt to a changing world. *UBC Business Review Journal*, 25-28.

Grass, E.M. (1990). *Employment and production: the mature stage in the life-cycle of a sawmill*. Unpublished PhD thesis, Department of Geography, Simon Fraser University.

Harvey, D. (1989). *The condition of post-modernity: An inquiry into the nature of cultural change*. Oxford: Blackwell.

Hay, E.M. (1993). *Recession and restructuring in Port Alberni: Corporate, household and community strategies*. Unpublished MA thesis, Department of Geography, Simon Fraser University.

Hayter, R. (1987). Technology and jobs: Innovation policy in British Columbia and the forest product sector. In K. Chapman and G. Humphry (eds.), *Technical change, unemployment and spatial policy* (pp. 215-32). Oxford: Basil Blackwell.

Hayter, R. (1997). High performance organizations and employment flexibility: A case study of *in situ* change at the Powell River paper mill, 1980-94. *The Canadian Geographer, 41*, 26-40.

Hayter, R., and Barnes, T. (1993). Labour market segmentation, labour flexibility, and recession: A British Columbian case study. *Environment and Planning C: Government and Policy, 10*, 333-53.

Hayter, R., Grass, E.M., and Barnes, T.J. (1994). Labour flexibility: A tale of two mills. *Tijdschriff voor Economische en Sociale Geografie, 85*, 25-38.

Hayter, R., and Holmes, J. (1994). Recession and restructuring at Powell River 1980-94: Employment and employment relations in transition. *Discussion Paper No. 28*, Department of Geography, Simon Fraser University.

Holmes, J., and Hayter, R. (1993). Recent restructuring in the Canadian pulp and paper industry. *Discussion Paper No. 26*, Department of Geography, Simon Fraser University.

Johnson, A., McBride, S., and Smith, P. (eds.) (1994). *Continuities and discontinuities: The political economy of social welfare and labour market policy in Canada.* Toronto: University of Toronto Press.

Klopfer, A. (1995). *High technology industries in BC: The agenda for growth, a discussion paper.* Victoria: Victoria Science Council of British Columbia

Lipietz, A. (1996). The post-Fordist world. Paper presented at the Centre for Human Settlements, University of British Columbia, July 13, 1996.

Lucas, R.A. (1971). *Minetown, milltown, railtown: Life in Canadian communities of single industry.* Toronto: University of Toronto Press.

Malecki, E.J. (1986). Technological imperatives and corporate strategy. In A.J. Scott and M. Storper (eds.), *Production, work, territory* (pp. 67-89). London: Allen and Unwin.

Massey, D. (1984). *Spatial divisions of labour.* London: MacMillan.

Marshall, R., and Tucker, M. (1992). *Thinking for a living: Work skills and the future of the American economy.* New York: Basic Books.

Ocran, A.C. (1995). *Industrial homeworking and employment standards: A community approach.* Unpublished paper, Department of Geography, University of British Columbia.

Rauch, U. (1996). *The social construction of skills: Working knowledge of garment workers in a Vancouver clothing factory.* Unpublished PhD thesis, Department of Anthropology and Sociology, University of British Columbia.

Randall, J.E., and Ironside, R.G. (1996). Communities on the edge: An economic geography of resource-dependent communities in Canada. *The Canadian Geographer*, 40, 17-35.

Reed, M.G., and Gill, A. (1997). Community economic development in a rapid growth setting: A case study of Squamish, B.C. In T.J. Barnes and R. Hayter (eds.), *Troubles in the rainforest: British Collumbia's forest economy in transition* (pp. 263-285). Victoria, B.C.: Western Geographical Press.

Rees, K.G. (1993). *Flexible specialization and the case of the remanufacturing industry in the Lower Mainland of British Columbia.* Unpublished MA thesis, Department of Geography, Simon Fraser University.

Rees, K.G., and Hayter, R. (forthcoming). Flexible specialization, uncertainty and the firm: enterprise strategies in the wood remanufacturing industry of the Vancouver metropolitan area, British Columbia. *The Canadian Geographer.*

Savoie, D. (1986). *Regional economic development: Canada's search for solutions.* Toronto: University of Toronto Press.

Scott, A.J. (1988). Flexible production systems in regional development. The rise of new industrial spaces in North America and Western Europe. *International Journal of Regional and Urban Research*, 15, 130-54.

Sjoholt, S. (1987). New trends in promoting regional development in local communities in Norway. In H. Muegge and W. Stohr (eds.), *International economic restructuring and the regional community* (pp. 277-93). Aldershot: Avebury Press.

Stanton, M. (1989). *Social and economic restructuring in the forest products sector: A case study of Chemainus, B.C.* Unpublished MA thesis, Department of Geography, UBC.

Streeck, W. (1989). Skills and the limits of neo-liberalism: The enterprise of the future as a place of learning. *Work, Employment and Society*, 3, 89-104.

Woodbridge, Reed Associates (1984). *Secondary manufactured wood products in B.C.* Victoria: Ministry of Forests.

Zukin, S. (1991). *Landscapes of power: From Detroit to Disney World.* Berkeley: University of California Press.

SECTION III
Community in Transition

On Distance, Power and (Indirectly) the Forest Industry*

R. Cole Harris

Department of Geography, University of British Columbia

* This essay draws upon material first published in *The Resettlement of British Columbia: Essays on Colonialism and Geographical Change* (UBC Press, 1997)

As much as anyone of his time, Harold Innis understood that systems of transportation and communication are not politically and culturally innocent. His early studies around the problem of distance in the Canadian economy and his later studies, surveying long reaches of time, of the biases inherent in different media of communication, all showed the closeness of the relationships between the means of overcoming distance and the cultural and political control of space.[1] Later theorists of social power have largely agreed.[2] Canada, as Innis understood it, grew out of the spatial economy of the fur trade, reinforced by railway and telegraph. He knew that these instruments of state formation were also, from another perspective, agents of colonialism and of cultural destruction, but did not pursue this side of the equation. His more central preoccupation was the Canadian question—the creation and continuing presence of Canada as a distinct North American society. Over the years this has been the author's preoccupation as well, but this chapter broaches the issues that Innis did not explore in British Columbia, a region that, by and large, he did not handle very well.[3]

A dialectic of creation and destruction surrounds any innovation in technologies of transportation and communication. When these innovations are imposed on territories long occupied by societies that have not contributed to the evolution of the technologies in question, then they become a central part of the colonial enterprise, means by which alien peoples take over land and remake it to their purposes. So it has been in British Columbia where the struggle to overcome distance can be thought of as a heroic story of progress in a new land, or as a sad reflection of ongoing and, apparently, unstoppable theft. No wonder that Native blockades of roads and railway lines provoke intense, conflicting emotions.

Broadly, the challenge of distance in British Columbia has been both external and internal. Externally, British Columbia was as removed from the centres of the modern world as any part of the globe; internally, its convoluted terrain and lack of prior transportation at all suited to the needs of modern societies posed an ongoing problem of distance that even today is very partially resolved. This chapter deals briefly with the collapse of external distance, says somewhat more about the challenge of internal distance, and then reflects on the implications of the changes described. The focus is not on the forest industry per se, but on the developments that permitted that industry to flourish on what had long been Native land.

EXTERNAL DISTANCE

For people of European background the northwest coast of North America at the end of the 18th century was as inaccessible as any mid-latitude coast on earth. Even in 1860, J.D. Pemberton, Surveyor General of Vancouver Island and British Columbia, estimated that the 17,000 mile trip from London ("the longest that can be taken from England to any known port rounding either cape") took almost five months.[4] The transcontinental connection was no easier. Express brigades carried letters and accounts (by canoe, pack horse, and river boat) from the lower Columbia to York Factory in three months. Despatches written in London in March or April were taken by ship to Hudson Bay and reached the lower Columbia at the end of October (Figure 12.1). Replies would leave the lower Columbia, headed east, late the following March.

Then, just before mid-century, the California gold rush provided the impetus for a series of drastic time-space compressions that allowed the outside world to engage British Columbia much more comprehensively. In the early 1850s American clipper ships and improved navigation to take advantage of winds and currents[5] cut average sailing time from New York to San Francisco to about 130 days, and the fastest sailings to 90 days. In 1855 a railway opened across the isthmus, making the 50-mile trip in five hours. In 1860 Pemberton calculated it this way: London to New York to Aspinwall (eastern end of Panama railway), 16 to 20 days; Panama to San Francisco, about 14 days; San Francisco to Vancouver Island, 4 to 5 days; total travelling time from London, England, to southern Vancouver Island, allowing for slight delays at New York and San Francisco, about 45 days (Figure 12.1). The trip could be made even more quickly by taking the train from New York to St. Louis and the express stage (22 days) from there to San Francisco.[6]

In 1865 a transcontinental telegraph reached New Westminster via San Francisco; shortly thereafter the turn around time for express communications between British Columbia and London was less than a week. The railway, completed to Port Moody at the head of Burrard Inlet in November 1885 and moved a few miles west to Vancouver and opened to passenger traffic before the end of 1886, was not much slower. With the opening of the Canadian Pacific, the distance between Vancouver and Montreal was calculated in hours (137); through passenger trains ran six days a week. Even in the mountains they maintained an average speed of 35 mph. The railway integrated British Columbia in a new transcontinental state and marked the final 19th century stage of the extraordinary sequence of time-space compressions that is shown in Figure 12.1.

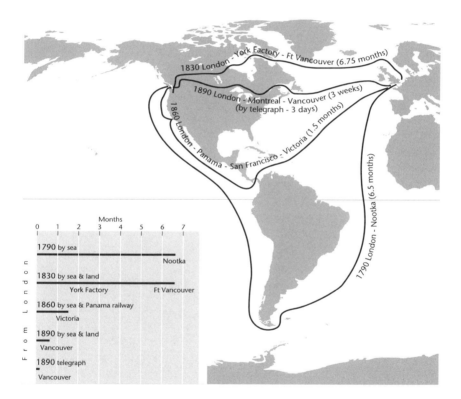

Figure 12.1   Time-space compression between London and the Northwest Coast, 1790-1890

The nineteenth-century assault on space and time was most dramatic at the margins of the world system. In a few years British Columbia's external connections had been transformed: by 1890 telegraphs carried compact, high-value information in and out of the province in a matter of hours; and mail arrived from eastern Canada in about a week, from Britain in two weeks, and from China and Japan in a month.[7] Bulk freight moved far more readily than ever before. Such changes expanded British Columbia's connections with the world economy, situated it squarely within the circuitry of a global empire, and underlay the introduction of government and the emergence of settler society. The colony of Vancouver Island coincided with improved communications with London following the California gold rush. British Columbia became a province of Canada on the (promised) overland strength of telegraph and railway. As shipping costs decline, industrial fish canning, sawmilling, and coal mining began on the coast; and by 1890 the first hard-rock mines were shipping high-grade silver ore from the Kootenays. Cattle ranching, an extensive land use on low-value land at the very edge of the world economy, took over much of the dry belt. Immigrants were attracted by prospects that apparently could be reached without undue risk or loss of contact with home. Their coming accelerated the colonial appropriation of land. Yet within British Columbia there were sharp geographical variations in the extent of connection with the outside world. The few towns on southeastern Vancouver Island and in the lower Fraser Valley and, by 1890, stations along the main line of the Canadian Pacific Railway (CPR) were the primary points of connection beyond which, more or less rapidly, the outside world dropped away.

INTERNAL DISTANCE

As late as 1870 only a few elements of modern transportation were superimposed on a huge, otherwise-unorganized land (Figure 12.2). There were small steamers on a few rivers and, in 1870, some 25 post offices. This hardly changed during the 1870s. In the 1880s, the CPR and another telegraph line arrived, but they were lines without thickness. Neither, in themselves, had the capacity to occupy a territory, certainly not one without the underlying networks of transportation and communication and the settled population on which, in Europe and eastern North America, 19th century technologies were superimposed. In British Columbia, outliers of civilization seemed suspended in wilderness. When the railway arrived it suddenly became physically easier (though not cheaper) to reach Montreal from a point along the line than many places only a few miles away.

Modern distance-diminishing technologies were imposed very incompletely, and with great difficulty, on this vast, angular land. Yet, because the white control and use of it required them, the effort to introduce them was enormous. The most aggressive impetus, by far, came from capital as it identified resources and sought to connect them to markets. The state did what it could to encourage capital, extend its own territorial reach, and, after a time, service a growing settler population. Settlers themselves built trails and roads here and there but, by and large, relied on a communicative grid imposed by capital and government. The main channels of this grid were railways, steamer runs, and, eventually, a few roads, whereas the interstices tended to be reached by more vernacular, western North American methods.

In the southern interior, and eventually in much of the central interior as well, networks of transportation and communication developed around railways. Operating on precise schedules, they emphasized clock time, and they

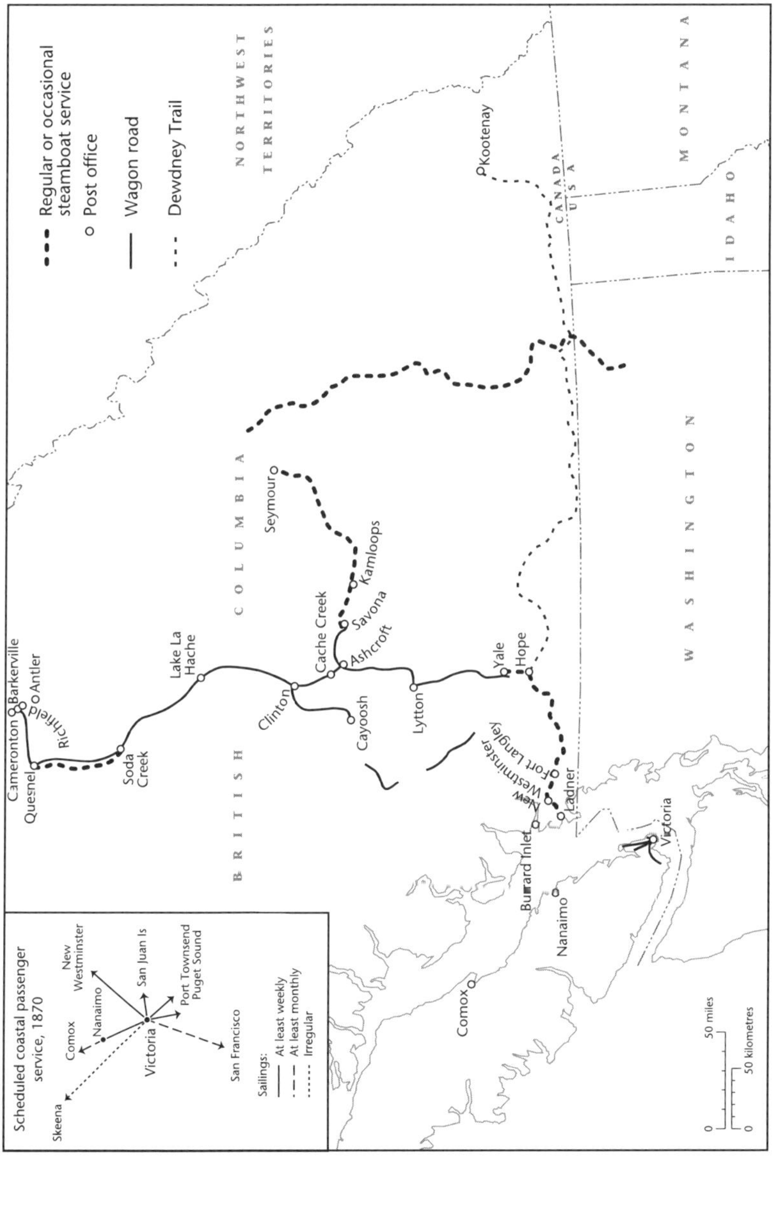

Figure 12.2 New transportation system in southern British Columbia, 1870

substantially nullified three age-old constraints on land transportation: night, inclement weather, and seasonal climates. Their artificiality gave them their power, allowing steam engines to haul by day or night, summer or winter, in a great variety of terrains.[8] But they were expensive to build, especially in mountainous terrain, and without lateral extent. In British Columbia they rarely ran through what Europeans would have called countryside.

In the southern interior, transcontinental railways—the CPR in the north, and the Northern Pacific and, from 1893, the Great Northern south of the border—bounded the transportation system (Figure 12.3).[9] The Great Northern was built expressly to drain the Canadian trade southward, and the principal mining activity in British Columbia at the end of the nineteenth century lay between the Great Northern and the CPR. Vying for this trade, both companies built feeder lines into the north-south-tending valleys between them. At huge expense, and over a period of some 20 years, the CPR also opened another line across southern British Columbia and the grain of the topography (the Kettle Valley Line), completing it in 1916 when the principal mining excitements in southern British Columbia were over. Two new transcontinental lines, crossing the Rockies well north of the CPR, opened as the first World War began: the Grand Trunk Pacific and the Canadian Northern. The provincial government built ineffectively northward from the head of Howe Sound near Vancouver, intending but failing to reach the Grand Trunk. Figure 12.4 shows the railways in British Columbia (excluding logging railways) in 1917 when their construction stopped until well after World War II: in total, some 3000 miles of track, many of them enormously costly, the heart of the inland transportation system.

River and lakes steamboats (Figures 12.3 and 12.4) filled some of the gaps in the railway system with a flexible, inexpensive technology that had evolved in the interior waterways of North America.[10] Small steamboats could operate in two feet of water, and nose ashore almost anywhere. In most of the larger valleys of southern British Columbia steamboats provided the first steam power, and some of them remained links in railway systems as late as World War II, transporting freight cars as well as passengers and local freight. Because of transshipment costs, they rarely competed for long with railways over the same routes, and were discontinued as the railway network expanded.

Roads, the only other public carriers of inland passengers and freight, were few and most of the province was not served by them (Figure 12.5). There were no roads along the coast north of Georgia Strait, none north of Barkerville, or in the Coast Mountains; even in the Kootenay mining regions, the most populous parts of the interior, there were no through roads. Surviving parts of the Cariboo Wagon Road, built from 1862 to 1865 and originally 18 feet wide, cambered, and well-graded and bridged, were still the best roads in the province. In 1900, the characteristic road in British Columbia was a dirt track 12 to 14 feet wide. As the number of motor vehicles increased, the need grew for a

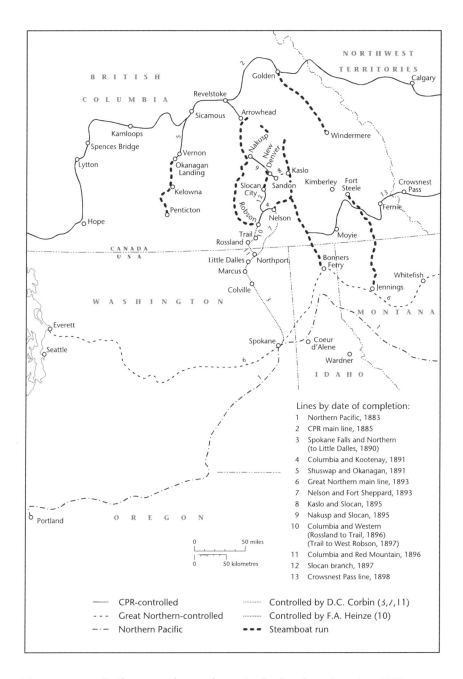

Figure 12.3　Railways and steamboats in the Southern Interior, 1898

different, much-expanded road system. By 1930, the provincial Department of
Public Works had replaced most of the wagon bridges, opened a road through
the Fraser Canyon (the first road link between coast and interior since CPR
construction had destroyed sections of the Cariboo Wagon Road), and had three
roads through passes in the Rockies (Figure 12.6). These dusty, washboard,
dirt or gravel roads were often virtually impassable in spring and fall; the new

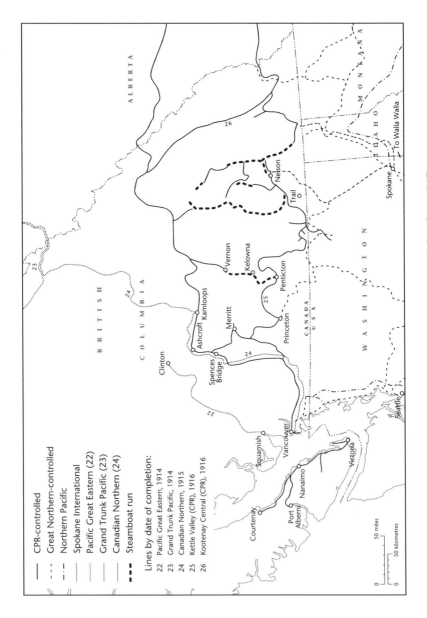

Figure 12.4   Railways and steamboats in Southern British Columbia, 1917

road in the Fraser Canyon was two feet narrower than its predecessor. The system had expanded, but remained meagre. In 1930 there was only one road from the interior to the coast. The road trip within Canada from Vancouver to Calgary, while possible, was circuitous and slow, as were most provincial roads. The far easier route east was south, then through adjacent American states.

Figure 12.5   Principal roads, 1903

Coastal waters accepted modern technologies of transportation far more easily. Throughout the 19th century, most coastal steamers were wood- or coal-burning side- or stern-wheelers, but by World War I steel hulls and propellers had largely taken over, and some steamers had been converted to oil. Overall, the coastal fleet was a motley of vessels of different sizes and capabilities that linked the principal towns and camps along a fjorded coast from Alaska to Puget Sound for a fraction of the cost of the emerging network of overland transportation.

Figure 12.7 shows scheduled passenger and freight sailings along the coast in 1890, 1901, and 1921. The cartograms show an expanding system increasingly focused on Vancouver, and suggest something of the services various companies provided. Union Steamships relied on small, multi-purpose freighters that supplied the logging camps, canneries, mines, and few farms scattered along the inside passage between Vancouver Island and the mainland; and

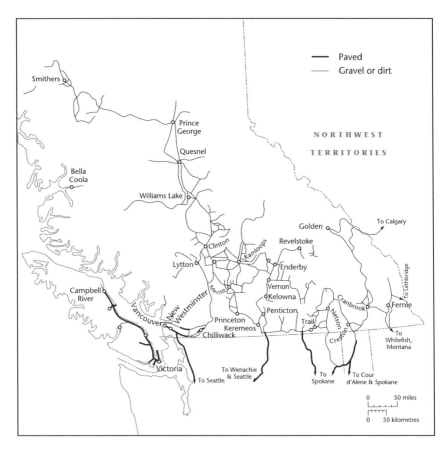

Figure 12.6   Principal roads, 1930

made weekly runs farther north to Bella Coola, Rivers Inlet, and the Skeena. Like the river boats, Union ships would put in to isolated settlements on demand.[11] The Canadian Pacific Navigation Company made longer, less flexible runs to the north coast and Alaska, controlled most of the passenger traffic across southern Georgia Strait between Vancouver, Victoria, and Nanaimo, and provided weekly or monthly service to the West Coast of Vancouver Island from Victoria. The Grand Trunk, by 1921 part of the Canadian National

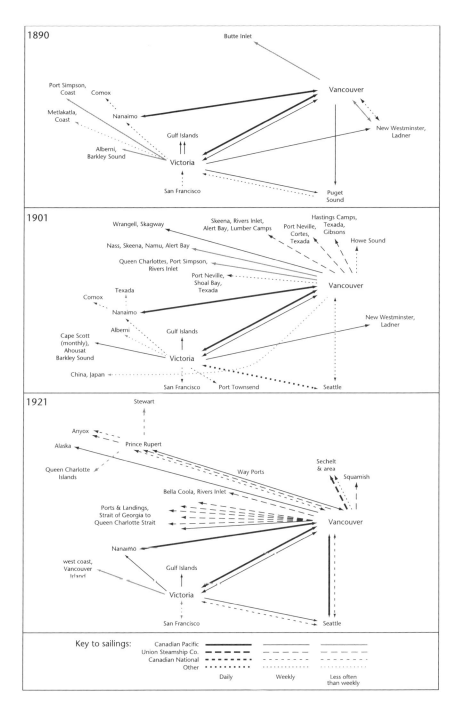

Figure 12.7   Scheduled coastal sailings, 1890, 1901, 1921

System, provided fast runs between Prince Rupert, Vancouver, Victoria, and
Seattle, and a more local service, analogous to Union Steamships', out of Prince
Rupert.

These different transportation systems all carried the public mail, the principal form of long-distance communication throughout these years. A post
office, a crucial commercial and personal link with the outside world, became
part of any new, fairly permanent settlement. Figure 12.8, showing the distribution of post offices in 1890, is a fair summary of the distribution of immigrant

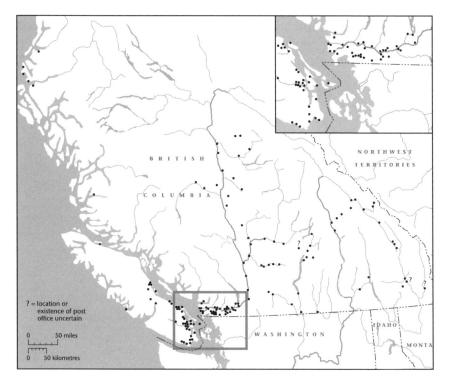

Figure 12.8   Post offices, 1890

settlement: a good many post offices on southeastern Vancouver Island and in
the Lower Fraser Valley, some along the route to the Cariboo, some along the
CPR, some in the new ranching or mixed farming settlements south of the
CPR, few anywhere else.[12] As immigrant settlement expanded, post offices
became more numerous: Figure 12.9 shows their distribution in 1915 in the
Kootenays, then the most populous part of the interior. Almost wherever there
were settlers, there was a post office nearby, a measure both of a high degree of
access to the outside world and of local limitations on personal mobility and
mail deliveries.

Telegraph, railway, road, steamer, and mail service comprised the main elements of the new communicative environment that reconfigured British Columbia before World War II. By any measure, their introduction was an impressive achievement—an enormous investment of capital, labour, and lives to reshape spatial relations and facilitate modernity. Yet in themselves they were an outline without much capacity to reach beyond a few narrow, introduced corridors of modern connectivity. Steamers connected points along the coast, but behind that coast, steamer whistles quickly faded into tangled forests and mountainsides. Along most of the 3,000 miles of railway the situation

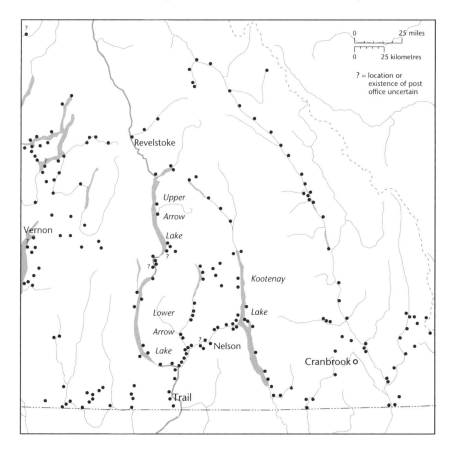

Figure 12.9   Kootenay post offices, 1915

was much the same. Essentially there were not enough roads, not the intricate miles of local tracks that in any older agricultural society had once taken carts and wagons and later would be improved to take automobiles and trucks. The interior of British Columbia had essentially one road in 1870, and although the road system had expanded greatly, it was still minimal 70 years later. Without prior roads, new corridors of transportation would not, automatically, have any lateral effect. And yet the resources immigrants identified in British Columbia lay in the spaces beyond the transportation corridors, often high on mountains or tucked away in inaccessible valleys. Here was a more local challenge of distance.

Martin Allerdale Grainger's novel *Woodsmen of the West*, set in an isolated coastal logging camp before World War I, is a British Columbian parable that turns around this problem. The steamer Cassiar (Union Steamships) brought a tenderfoot English logger from Vancouver to Port Browning, near the mouth of Coola (Knight) Inlet 175 miles north of Vancouver. His destination, however, was a logging camp; a small, thoroughly marginal operation 70 miles beyond Port Browning and the Cassiar up Coola Inlet. Much of the novel deals with this tenuous connection: a succession of tiny, makeshift steamers (the best of them a "god-forsaken looking tub"), their propellers or rudders dropping off, their boilers bursting, their rotten planks leaking. There were night runs up dark channels. There were groundings and sinkings. Then, because the camp was eventually steamerless, its connection to Port Bowering became a solitary 70 mile winter row in an open, leaking, 18 foot boat. At the logging operation, camp and donkey engine were on rafts, the "donk" was "jumped" from raft to shore, winched up-slope, it cable hauled 1,000 feet up-slope. Even jack screws "coaxed" logs to the sea. There was a 400 foot boom for towing logs. Everything was bent to "getting out logs." The owner of this haywire operation, as driven as Captain Ahab, identified with the machine that moved his logs: "I and the donk." His is the struggle to master nature and distance beyond the transportation corridors and at the very edge of the world economy.

In the mountains of southeastern British Columbia (the Kootenays), Grainger's tale could be told around prospecting and mining. Prospecting took place almost anywhere, but "proving up" a prospect required a horse trail, built by hand usually with enormous labour across difficult terrain. Mining would not begin until there was a railway or steamboat nearby—the impetus for the expansion of both in the Kootenays. The mines, however, were in the mountains and at first, when only the highest grade ore was shipped, the gap between them and steam power in the valleys was filled by pack trains or by rawhiding (using horses to pull ore wrapped in rawhide down mountain trails in winter). More developed mines required better connections: usually an overhead tram comprising towers, cables, and ore buckets built by an American contractor familiar with such systems; or a short, switch-backed wagon road. At the junction of tram or wagon road and railway or steamer run there was a mill, another link in the transportation system, through which ore was gravity fed and concentrated, then loaded into box cars. In the coastal forest the focus was on moving logs. When they were hauled by oxen over greased, corduroy roads, logging could not take place more than two or three miles from tidewater. With logging railways it could move well up valleys, while the donkey engine and high lead logging allowed it some distance up the valley sides. The reach of a donkey engine could be doubled or tripled by cold decking, that is by winching the donkey itself to the end of its cable, and building a platform (cold deck) from which its cable could reach another 1,000 feet. Local solutions were numerous. In some operations flumes transported shake bolts and (in the interior) logs. Some used steam tractors or cog railways on steep slopes. By the late 1920s the bulldozer was being used to build logging roads and the truck to haul logs, a combination that would soon take logging far beyond the range of the logging railways and, beginning in the 1930s, would close them down.[13] The salmon fishery, too, expanded its reach: from a gillnet river fishery that used flat-bottomed skiffs or Native canoes and targeted sockeye, to a much more wide-ranging, predominantly deep-sea fishery that took all varieties of salmon and used gas-powered fishing boats, tenders, and various nets and trolls.[14]

Farming, on the other hand, required roads, both on and off the farm. Off-farm roads in the transportation corridors were usually short, perhaps no more than a few hundred yards from a farm on a post-glacial terrace to a steamer landing on a lake. Where patches of farmland were broad enough to permit some density of farms, as in the Lower Fraser Valley or the Peace River, networks of rural roads emerged, although this was uncommon in British Columbia because agricultural land was so limited. Many farms were also connected to the uplands behind them. In the dry belt, creeks were dammed, and flumes, ditches, or wire-bound wooden pipes brought irrigation water, occasionally from dams as much as 30 or 40 miles away. Cowboys and shepherds drove cattle or sheep from bottomland winter pastures to high summer ranges, in the process occupying these ranges for particular ranches.

These local struggles with distance embodied much of the creative, practical ingenuity in early modern British Columbia, and created much of the texture of its changing human geography. Allowing the province's primary resource economies to expand well beyond the transportation corridors, they put flesh on a few bones and, by enabling newcomers to use ever more land, considerably filled in the modern map of British Columbia. The geographical transformation was huge.

DISTANCE AND POWER

Here and there these developments drew on Native precedent, but for the most part they did not. Immigrants considered Native ways irrelevant. They treated British Columbia as a *tabula rasa*, and constructed their own systems of transportation and communication essentially from scratch. These systems were among their principal achievements, the framework on which the province's changing human geography was hung. As such, they were enormously powerful, for they shaped many of the spaces in which British Columbians lived and, in so doing, many of their social relations.

The relationship between empire and the media of transportation and communication has been widely explored, in this country first and most suggestively by Harold Innis in his studies of the CPR, the fur trade, and the bias of communications. Railways, steamers, and telegraphs were admirable tools of imperialism, incorporating into empires territories won by the overwhelming superiority of European firearms.[15] As much as troops on the ground, a nineteenth century empire depended on the capacity of interlocking networks of rails, steamship runs, wires, and mails to overcome distance. As space collapsed, territory awaited the powerful, and the rhetoric of expansion became more and more self-evident. Strong states, it was held, would inevitably expand, and should absorb weaker, less civilized societies.[16] The question was not of the right to expand, which was taken for granted, but of which strong state would expand the most—the impetus for the rush of European states to carve up Africa. In this climate, geopolitical thinking flourished, and British Columbia was not detached from it. The CPR, for example, consolidated a transcontinental state, and enabled central Canada to reach quickly outward, as it did by both railway and telegraph to put down Métis rebellion on the prairies. It was also a link between Britain and the Orient; together with the empress ships that sailed out of Vancouver and that, by prior agreement, the Admiralty could commandeer at any time and convert to troop carriers, it provided an alternative route to India should the Suez Canal be blocked.[17] The Grand Trunk Pacific, built with millions of British pounds, was a final imperial northwest passage in a railway age.

Within British Columbia, imperial visions merged with the realities of colonialism. As distanced diminished, newcomers were able to possess the "wilderness" more comprehensively. Edward Said has suggested, in his enormously influential book *Orientalism*, that the construction of the Suez Canal "destroyed the Orient's distance, its cloistered intimacy away from the West," and that, unsheltered, the West could now possess it. "After de Lesseps no one could speak of the Orient as belonging to another world . . . . There was only "our" world . . . ."[18] So, too, in British Columbia, as its distances were undermined. Claiming political control of a territory was an act of colonialism, coming to know it was often another, but using it was far more intrusive than either. Improvements in transportation and communication enabled the world economy to use British Columbian space not through Native intermediaries, as during the fur trade, but by distributing Western technologies, labour, and settlers across the land. They allowed the state to greatly expand its reach, providing the channels by which it collected information, distributed regulation, and imposed order. Although it would not quickly be possible in British Columbia to extend the state uniformly over its territory—which the sociologist Anthony Giddens holds to be one of the basic characteristics of modern societies—the thrust, facilitated by the decline of distance, was in this direction. From a Native perspective, white territorial claims, place names on maps, and exploration and survey parties were being followed by a far more tangible form of colonialism: white workers, settlers, and their machines using Native land for their own purposes. As long as Natives had been able to hunt, fish, and gather in their former territories, the small areas the Reserve Commissioners laid out for them had little meaning, but the implications of the reserves, and the exclusions they entailed, became evermore apparent as non-Natives occupied and used the surrounding land. A logging operation in a mountain valley previously used for seasonal hunting marked a huge transformation of land use and power. For those experiencing it, colonialism was enacted locally, on the ground.

Viewed in this light, the systems of transportation and communication that spread into British Columbia were the capillaries of colonial appropriation. They allowed non-Natives into the land, not as explorers, visitors, or passers-through, but as users and settlers. Coupled with a regime of property rights validated by the state, they effected a vast transfer of local power from indigenous peoples to immigrants. They did so not as the result of any one event, but incrementally, as a flume was built here, a corduroy road laid out there, a donkey engine brought into the bush. This pervasive geographical expansion, superimposed on a depopulated land and backed by the immigrants' cultural confidence, technological superiority, and force of arms, was itself a diffused form of power that prized open more and more nooks and crannies of an alien land and civilized them in terms a modern immigrant population could understand. The influential French social philosopher Michel Foucault has insisted upon the relationship between social power and the configuration of space, but whereas Foucault's examples turn around prisons, asylums, and reformatories, in British Columbia the prime example is the land itself as it was being reconfigured into new patterns of appropriation and social control. Colonialism and colonization were about the control of land; land use itself defined new rights, exclusions, and patterns of dominance; and strategies for the effective control of land operationalized colonial rhetoric and discourses. In British Columbia immigrants were preoccupied with the challenges and opportunities of a remote land into which global imbalances of power and the British empire had incongruously brought them, and as they diminished the problems of distance and gained access to the land, they relegated the former population ever more to the sidelines. A Foucauldian analysis of the decentred strategies of disciplinary power is not invalidated in such colonial settings, but it is shifted—to be come more preoccupied with land and the disciplinary power associated with reaching and possessing it.

There is still another side of the colonial question: the extent to which new systems of transportation and communication were themselves agents of Native cultural change. This is a matter awaiting exploration, but it is worth bearing in mind the conclusion of Jean and John Comaroff (with respect to Protestant missions in South Africa), that colonized peoples often "reject the message of the colonizers, and yet are powerfully and profoundly affected by its media." The very meeting of different ways, they suggest, had the capacity to redefine "the taken-for-granted surfaces" of everyday worlds. In the missionaries' implicit assumptions, rather than their explicit evangelical narratives, lay the "hegemonic forms" that shaped "colonial subjects."[19] If a steamboat or a railway were less articulate than a missionary, they perhaps took as much for granted: clock time, linear distance, rational empiricism, a revised geographical relationship with space and with other peoples, an instant relativity that situated the local in the world. A railway that ran past a Native village was more than an intrusive symbol of white power; it redefined the "surfaces" of life in that place, making people there more self-conscious, situating them within a global, rationalized civilization, taking away their local integrity. A talking wire (the telegraph), initially perceived as another form of white spirit power, eventually led to a wholly other way of thinking about the world.[20] The new media of transportation and communication powerfully unbalanced white-Native interactions; they appear to have been powerful colonizers of Native consciousness.

From the newcomers' perspective most of this was invisible. The land was wilderness awaiting development. The few legacies of a native past were irrelevant to an immigrant future; British Columbia was a beginning. In terms of the media of transportation and communication this meant not only that systems were constructed *de novo*, but also that they were constructed without interference from the past. Liberated from the past, they faced an open, untrammelled relationship with the future, a relationship affected by the terrain through which they would pass, but not by prior human geographies. Immigrant British Columbians knew this. The location of their future was still very much up in the air, an uncertainty that was the source of the strident boosterism with which they promoted this or that settlement. Transportation schemes frequently underpinned such boosterism, also for obvious reasons. Transportation would provide the links with other places; in good part settlements would grow in proportion to the strength of such links; and a railway (or other form of modern transportation) was often the difference between an efficient connection and none at all. The transportation system, in short, was seen to have enormous power, not to complement or supplement, but to create; and there was abundant evidence at hand that it did.

Victoria, the first town, grew out of the Hudson's Bay Company's main Pacific depot, moved from the lower Columbia to southern Vancouver Island in 1843 to an accessible harbour for the company's growing Pacific trade.[21] New Westminster emerged during the gold rush as a classic gateway town, a river-mouth, steamboat port on the main route to the diggings. Vancouver, which superseded New Westminster, was a creation of the railway, and its spectacular growth to metropolitan dominance reflected its location at the junction of the main corridors of transportation affecting the province: railway routes east, coastwise steamer traffic, and deep-sea shipping.[22] At no other point in a network dominated by a few channels was there anything like equivalent access to goods and information. By the time Prince Rupert became the Pacific port for the Grand Trunk Pacific, the pattern of urban dominance was already set; Prince Rupert could not begin to compete with the many linkage advantages, both within the city and beyond, that already had accrued to Vancouver.

Another category of towns, all of them small in early British Columbia, emerged at points where local resources and access to distant markets were both available. Nanaimo, the first of them, was a colliery town on the coast, accessible to shipping. Across Georgia Strait and later, Britannia was a company town beside a concentrating mill and at the junction of two transportation systems: a tram bringing copper ore to the mill from mountainside mines, and ships loading concentrates. At Hedley in the Similkameen Valley, and at many other small hard-rock mining towns, the pattern was essentially the same. The closer the town to the resource base it depended on, the more ephemeral it was likely to be: Phoenix, atop a copper mine in the Boundary Country, became an open pit; Anyox, another copper town on the coast north of Prince Rupert, closed down with its mine. Sandon, a silver-lead town in the Kootenays, faded away as its small mines closed; Kaslo, on the other hand, a gateway to the Sandon mines via a narrow gauge railway from Kootenay Lake, survived. Small developments that provided short-term employment remained camps—perhaps an adit, rock dump, a few mine buildings, and bunk-house accommodation for men high on a mountain. A logging railway might produce a sawmill town (like Chemainus), but logging itself produced camps. The camps of the salmon fishery were seasonally occupied canneries, located close to the resource and accessible to steamers; at the mouth of the Fraser there were enough canneries bunched together to created a small town (Steveston). In the few places where there were dispersed farming populations, as in the lower Fraser Valley, the Okanagan, and the Peace River, small towns emerged as rural service centres. Some towns combined several of these functions. All of these different towns and camps were at advantageous locations in a new matrix of transportation and communication that was a necessary condition for their existence.

Regional economies and land uses were also being reworked as distances declined and the reach of the international market expanded. Here, as elsewhere, the market's influence was to commodify land and specialize land use. Trees began to be calculated in board feet, salmon in the number of 24 can packs. Clearcuts were products both of techniques of logging and hauling and of the market's unrestricted access to forests. Various forest tenures gave clear title, the Natives were on reserves, and, apart from their discounted, largely inaudible voices, there was no interference from the past or from resident populations with other agendas. The equation seemed greatly simplified: wilderness and market bound, as it were, by logging railways and donkey engines. Provincial land policies that allocated land for specialized purposes (forestry, mining, agriculture) provided institutional support for an international market economy that depended on the division of land and the functional integration of economic space as much as on the division of labour. In older societies regional specializations tended to be imposed slowly in the face of a good deal of resistance from prior, more regionally self-sufficient economies, but in British Columbia they were the abrupt accompaniment of the virtually simultaneous arrival of settlers, low-cost transportation, and the market. Therefore, ranching, logging, or mining quickly dominated particular regions; and at various scales heartland-hinterland relationships were quickly introduced. Economic specialization tended towards ecological simplifications—for example, as agriculture or, later, tree planting simplified prior ecologies. In these various restructurings was a whole new geographical order of things.

In short, introduced systems of transportation and communication incorporated British Columbia within the modernizing world, and created patterns of settlement and land use that bore many characteristic stamps of modernity. But modernity is itself a collection of shifting relationships, and relocation adjusted the complex of modern ways that reached British Columbia from the east. These adjustments were bound up with the particular flows of goods, people, and information into and within the province.

Most immigrant British Columbians could stay closely in touch with the places they had left. Letters and all types of printed materials could find them, keeping them in contact with distant families, prevailing opinion in home societies, and world news. Letters were especially awaited, and protracted correspondences, lifelines between distant worlds, were common. Newspapers and magazines could be sent from home, and books not locally available could be ordered. Mail order catalogues displayed current fashions and consumer goods, some of which were usually available in even the smallest general dry goods stores. Few came to British Columbia to escape the modern world, and those few hardly could, for a modern communicative environment was at hand with a vast capacity to distribute information. At the same time, immigrant and home societies communicated with each other out of different contexts, which over the years diminished their capacity for mutual understanding. A middle class English wife, prim, brave, and as unprepared as the British army officer-turned-orchardist she had recently married, lived surrounded by forest on a terrace clearing in the Windermere Valley in southeastern British Columbia, and corresponded with a mother and sister in a four-storey Georgian town house on High Street in Windsor, England. She knew their world but could not reproduce it; they could only imagine hers. A Chinese coal miner living in the spare male accommodation of Nanaimo's Chinatown received a scribe-written letter from his wife in a peasant village near Canton—man and wife living in vastly different settings thousands of miles apart. A pioneer wife learned about the latest domestic science at a local meeting of the Women's Institute and returned to a log cabin with a water barrel in a corner. A colonial secretary, writing from Downing Street, communicated a theoretical experience of empire to officials in British Columbia who somehow had to accommodate such theory and their sense of local realities. Texts, in short, travel more readily than contexts, and to the extent that both are required for communication, British Columbians were on their own.

More fundamentally, perhaps, British Columbia had reproduced neither space nor time as they were commonly understood in Europe. It appeared to have simplified space. This was partly because immigrants considered that they had left the past behind, an assumption that deeply implicit assumptions about the location of civilization and savagery and plummeting native numbers both encouraged. To take the past out of space was to eliminate most of its human texture, a virtually unimaginable subtraction even in a rapidly urbanizing, industrializing Britain, the heart of the modern world, but an obvious fact of immigrant life, apparently, in British Columbia. It was partly because immigrant activities tended to be spatially segregated by the uneven distribution of resources and the specializing tendencies of the international market. One region was dominated by the equipment, work routines, logistics, and, later, the subculture of logging; another was similarly dominated by mining; another by ranching. Regions changed as resources were depleted and technologies and markets evolved, and they variously overlapped, but large stretches of land tended to be dominated by a few particular strategies of resource extraction. The economy of British Columbia as a whole depended on a few resource industries, each with its characteristic, much repeated human geography. Vancouver's metropolitan dominance was another form of spatial specialization, in this case around flows of goods and information. Simplified space was also partly the creation of people who wanted to avoid complexity and relocated it when they could: by legislating reserves for Natives or by the constant pressure of racism that largely created Chinatowns and Japtowns.[23] As Edward Said has pointed out, people who are themselves dislocated and threatened by the unknown and the previously distant tend to fall back on

their own basic texts, in this case about the superiority of European civilization and the inferior otherness of the rest of the world.[24] Essentially, a relocated, simplified version of (loosely) "British" culture sought to contain the
unfamiliar complexities of its new situation.

This textual agenda and related spatial strategies, coupled with the denial
of a local past, and the territorial specializations inherent in international market economies and supportive government policies, encouraged new, simplified constructions of space. Men worked in camps, enclaves of capital and
labour largely abstracted from social contexts (other than ethnicity, itself an
abstraction) and relocated in wilderness. As there were often no other, the lines
of industrial transportation became those of social interaction. Men in mountainside bunk-houses rode the ore tramways to nights out in the bars below.
Men leaving the coastal logging camps or canneries caught the steamer to Vancouver; there was nowhere else to go. The drab toil of a work camp, the bright
lights of a city; the tough maleness of a work camp, the softer, civilized female-
ness of home: simple spatial dichotomies within simple constructions of space.[25]
Much of the interior dry belt was quickly known as cattle country and recognized as such in provincial land policy, but the very speed of such homogeneous regional identifications is a measure of the lack of perceived alternatives.
Even in Vancouver, rows of California bungalows emerged on the west side of
the city, their occupants white, English-speaking, middle class people who
lived, as much as possible, within networks linking others of their kind in
Vancouver and home societies in eastern Canada or Britain. A modernizing
British Columbia denied the many novel elements of complexity within it.

And it appeared to have jumbled time, displacing it, destroying its linear-
ity, mixing elements from the past like raisins in a pudding. There was little
continuous British Columbian time, rather, essentially, a present and its future.
The European past was relevant but distant; it contained the history of most of
the people who came to British Columbia but not of the place where they lived.
Yet artefacts from this geographically displaced past crept in. Settlers built
log cabins, dwellings not seen in Western Europe south of Scandinavia since
the medieval forests were cleared. Packhorse trails and railways intersected.
Wherever transportation costs were high, local labour and pre-industrial technologies were viable alternatives to imported manufactures—as when Hawai-
ians pit-sawed lumber at Fort Langley and the fort blacksmith made tools that
in Britain at the time were factory-produced. In such ways immigrants lived
with introduced anachronisms. And the ethnographers sought out "traditional"
Native cultures, their quest, one might suspect, for the original, uncivilized
Europe of the noble savage. In this light British Columbia appeared to contain
the beginning and the current end of Europe together with a few intervening
artefacts.

These brief and obviously preliminary reflections are perhaps relevant to Innis and the forest industry. Innis was a student of distance, and of the implications of different means of overcoming it. The forest industry has been, essentially, a particular means of dealing with distance. Because of the distribution of forests, the industry based on them has been spatially extensive. It, far more than any other activity in British Columbia, has marked the landscape and defined a relationship with nature. It has become a culture that is both cause and effect of systems of transportation and communication. As such, it runs squarely into other cultures, some of them, in this colonized place, deeply indigenous, and others part of the ongoing dialectic between progress and nature within western civilization. The trouble in the rainforest stems, fundamentally, from these confrontations.

ENDNOTES

[1]   Innis's main writing is collected and his thought summarized in: Drache, D. (ed.) (1995). *Staples, markets, and cultural change: Selected essays: Harold Innis*. Montreal and Kingston: McGill-Queen's University Press.

[2]   Among the most imposing of this literature are: Giddens, A. (1981). *A contemporary critique of historical materialism*, Volume 1. Berkeley: University of California Press, and Giddens, A. (1987). *The nation state and violence*. Berkeley: University of California Press; Harvey, D. (1989). *The condition of postmodernity: An enquiry into the origins of cultural change*. Oxford: Blackwell; and Foucault, M. (1977). *Discipline and punish: The birth of the prison*. London: Penguin, and Foucault, M. (1980). *Power/knowledge: Selected interviews and other writings, 1972-77*. Brighton: Harvester.

[3]   For example, see Innis, H.A. (1936). *Settlement and the mining frontier*. Toronto: University of Toronto Press. Innis amassed a lot of information, but did not understand the Kootenays, the principal regional focus of his attention, and frequently jumbled places and events.

[4]   Pemberton, J.D. (1860). *Facts and figures relating to Vancouver Island and British Columbia, showing what to expect and how to get there*. London, p. 84.

[5]   Lieutenant Matthew Fontain Maury's Explanations and Sailing Directions to Accompany the Wind and Current Charts, 7th ed., Philadelphia, 1855; cited in Rydell, R.A. (1952). *Cape Horn to the Pacific: The rise and decline of an ocean highway*. Berkeley: University of California Press, pp. 127ff.

[6]   Pemberton (op. cit.), pp. 86-91.

[7]   Innis, H.A. (1923). *A history of the Canadian Pacific Railway*. London and Toronto: University of Toronto Press, pp. 138-9.

[8]   Cronon, W. (1991). *Nature's metropolis: Chicago and the Great West*. New York: W.W. Norton, pp. 63-81.

[9]   For a somewhat fuller account of the evolving pattern of railways in southern British Columbia see my earlier essay: Harris, C. (1983). Moving amid the mountains, 1870-1930. *BC Studies*, 58, 3-39

[10]   Again, see Moving amid the mountains, 1870-1930; or, in more detail: Affleck, E.L. (1973). *Sternwheelers, sandbars and switchbacks.* Vancouver: Alexander Nicholls Press. There are many detailed local accounts, a good recent example of which is: Davies, D.L. (1994). Sternwheelers on the Thompson. In W. Norton and W. Schmidt (Eds.), *Reflections: Thompson Valley histories.* Kamloops: Plateau Press.

[11]   Rushton, G.A. (1974). *Whistle up the inlet: The Union Steamship story.* Vancouver: Douglas and McIntyre.

[12]   For a comprehensive list of B.C. post offices including date of establishment and closure: Topping, W. (1983). *A checklist of British Columbia post offices*, Vancouver: self-published. See also Melvin, G.H. (1962). *The post offices of British Columbia, 1858-1970.* Vernon, B.C.

[13]   There is a large, lively popular literature on these matters. I also recommend: White, R. (1980). *Land use, environment and social change: The shaping of Island County, Washington.* Seattle: University of Washington Press (particularly on logging with oxen); Haig-Brown, R. (1942). *Timber.* New York: Morrow (particularly on patterns of logging with logging railways and donkey engines); and Kahrer, G. (1988). *Logging and landscape change on the North Arm of Burrard Inlet, 1860s to 1930s*, MA thesis, University of British Columbia, Vancouver, B.C. (particularly on the conveyances and contraptions that brought logs and shake bolts off the North Shore mountains to mills around Burrard Inlet).

[14]   Among the considerable literature on the salmon fishery, the work most closely focused on these matters is: Higginbottom, E. (1988). *The changing geography of salmon canning in BC, 1870-1931*, MA thesis, Simon Fraser University, Burnaby, B.C.

[15]   Headrick, D.R. (1981). *The tools of empire: Technology and European imperialism in the nineteenth century.* New York: Oxford University Press; and Headrick, D.R. (1988). *The tentacles of progress: Technology transfer in the age of imperialism, 1850-1940.* New York: Oxford University Press.

[16]   Among the large literature on this topic see particularly: Kern, S. (1983). *The culture of time and space, 1880-1918*, Cambridge, MA: Harvard University Press (particularly chapters 8 and 9); and Said, E. (1979). *Orientalism.* New York: Vintage Books (particularly Chapter I, Parts I and II).

[17]   Innis, H.A. (1923). *A history of the Canadian Pacific Railway.* Toronto: University of Toronto Press; Hamilton, J.H. (1956). The 'All-Red Route, 1893-1953: A history of the trans-Pacific mail service between British Columbia, Australia, and New Zealand, *British Columbia Historical Quarterly*, XX(1 & 2).

[18]   Said, E. (1979). *Orientalism.* New York: Vintage Books, p. 92

[19]   Comaroff, J., and Comaroff, J. (1991). *Of revelation and revolution: Christianity, colonialism, and consciousness in South Africa.* Chicago: University of Chicago Press, pp. 309-314.

[20]   In some cases Natives adapted very quickly. In 1866, less than two years after the telegraph arrived, Nklka'pamx people sent a telegram in chinook to the Anglican missionary J.B. Good at Yale inviting him to establish a mission at Lytton. Good went, interpreting the telegram as a call from God. See Christophers, B. (1995). *Time, space and the people of God: Anglican colonialism in nineteenth century British Columbia*, M.A. thesis, University of British Columbia, Vancouver, B.C.

21 Mackie, R. (forthcoming). *The Hudson's Bay Company on the Pacific, 1821-1843*. Vancouver: UBC Press.

22 Consider the analogous relationship of St. Louis and Chicago, admirably described by: Cronon, W. (1991). *Nature's metropolis: Chicago and the Great West*. New York: W.W. Norton, pp. 295-309.

23 Anderson, K. (1991). *Vancouver's Chinatown: Racial discourse in Canada, 1875-1990*. Montreal: McGill-Queen's University Press.

24 Said (op.cit.), p. 93.

25 The work environment in British Columbia exacerbated, and provided an extreme example of, the growing, gendered disjunction between domestic and economic space that characterized industrializing urban societies in the 19th century.

Plate 20 (overleaf)   The Victoria Lumber & Manufacturing Company   ➤

Vancouver as a Control Centre for British Columbia's Resource Hinterland: Aspects of Linkage and Divergence in a Provincial Staple Economy*

13

Thomas A. Hutton

School of Community and Regional Planning
University of British Columbia

* This essay first appeared in *BC Studies*, Spring, 1997

INTRODUCTION AND CONTEXT

Sixty years ago Harold Innis presented the staple theory as a singularly powerful and evocative explanation of Canada's economic history and, more particularly, the crucial role of resource extraction in the development of Canada and its constituent regions. A recurrent feature of Canadian economic development over several centuries was the extraction and transportation of staples from hinterland regions to distant markets. As Innis observed, the ". . . result was that the Canadian economic structure had the peculiar characteristics of areas dependent on staples—especially weakness in other lines of development, dependence on highly industrialized areas for markets and for supplies of manufactured goods, and the dangers of fluctuations in the staple commodity" (Innis, 1933: 6). The dominant staple within the Canadian economy has followed a progression—from fish, to furs, minerals, forest products, grains, and so on—and of course transportation technology has evolved over time, but this classic portrayal of Canada as a staple economy remains cogent.

A cardinal feature of the Innisian staple model is the notion of a core-periphery structure of inter-regional linkages and relationships. At the broadest level, this implies a highly asymmetrical and dichotomous spatial framework, characterized by a dominant, industrialized and metropolitan "core," and a vast, underdeveloped "periphery" almost wholly dependent on the production and export of staple commodities. The core region enjoys a privileged position within staple economic regimes by virtue of its concentration of corporate management functions, transportation "gateway" role and entrepôt function, and growth of manufacturing impelled by what Innis called a kind of "cumulative industrialism" (*ibid.*). The core region could experience some measure of industrial development and diversification, owing to its larger local market, access to labour and capital, and agglomeration and urban economies—attributes notably lacking within the vast hinterland regions.

While a full comprehension of the dynamics of the core-periphery structure requires considerable elaboration, the asymmetrical nature of inter-regional relationships represents perhaps the essential feature. This asymmetry is largely defined by the "command and control" functions exercised by the dominant metropole over the periphery, by the qualitative imbalance in trade flows between core and periphery (and specifically the imperative of the latter to supply the former with resources), and by the consequently truncated nature of the hinterland economy mired in the well-known "staple trap." At the national level, Innis depicted Canada's extensive peripheral regions as largely dependent on supplying resources (minerals, forest products, energy) to the manufacturing and commercial heartland of Canada concentrated in the metropolitan Toronto and Montréal regions (and their industrial satellites). At the same time, such peripheral regions served as a captive market for end products from these industrial metropoles (Innis, 1933), thus producing boom and bust cycles.

This classic core-periphery structure is faithfully replicated in the case of British Columbia, within which the trajectory of economic development over the postwar period includes both the massive expansion of resource extraction (supported by large-scale public as well as corporate investments) within interior and peripheral regions, and the concomitant growth of Vancouver as provincial primate city and metropole. This latter role included large-scale resource processing industries in the city proper and its suburbs, and, more recently, the development of a downtown corporate complex of head office, financial, and producer services from which Vancouver exercises its provincial control functions, and which symbolizes the specific nature of Vancouver's primacy within a vast resource economy.

As central as these control functions are to Vancouver's positioning within British Columbia, we can reference a richer, more comprehensive and more nuanced set of economic and socio-economic linkages, interrelationships and interdependencies. These include, among others, *production linkages*, incorporating Vancouver's role as processing and secondary manufacturing centre for resource commodities; Vancouver's *strategic transportation role*, notably the export of some 40 to 50 million tons of staple commodities annually; *consumption interdependencies*, including, certainly, a considerable flow of staples from the interior regions to the consumer markets of the Lower Mainland; and *socio-cultural relationships*, including the influence of the staple economy and resource industries on the formation of class and community structure in Vancouver. Even this cursory outline points to the strength and diversity of core-periphery linkages between Vancouver and the rest of the province.

The pervasiveness and variety of these linkages notwithstanding, there is considerable evidence that dynamics operating at several spatial scales are now exerting pressures for change within the regional economies of British Columbia; that relationships between Vancouver and the rest of the province are subject to renegotiation and redefinition; and that consequently a reappraisal of the core-periphery structure in British Columbia is imperative. While it is premature to speak of Vancouver "de-coupling" from its resource hinterland (Davis, 1993), it is legitimate to posit a reconfigured core-periphery structure informed by a review of transformational or structural changes.

Earlier studies identified a "loosening of the bonds" between core and periphery within the context of a provincial staple economy (Ley and Hutton, 1987). It can be argued, however, that core-periphery divergence is accelerating in response to an interdependent galaxy of local and external dynamics. This chapter offers a fresh interpretation of Vancouver's relationship with British Columbia as staple economy in a context of ongoing structural change. This requires, in turn, an appreciation of macro-level economic forces, including the globalization of resource markets and industries (Marchak, 1995); diminution of British Columbia's extractive sector in the face of resource depletion, shifts in levels of external supply and demand, and changing social attitudes and policy decisions regarding allowable resource yields; and, finally, transformational change within Vancouver's industrial structure and labour force.

In terms of organization, each of five dimensions of core-periphery linkage—transportation, production, consumption, community structure, and specialized services—will be examined, and evidence for current or prospective change presented. Special attention will be given to an analysis of "command and control" functions exercised by the core region of Vancouver, because they are the crucial, defining feature of core-periphery relations. It will be argued that under any plausible scenario a confluence of exogenous and localized influences will in the aggregate promote divergence between core and periphery in British Columbia.

Transportation Linkages: Metropolis as "Commodity Export Gateway"

Vancouver owes its origin to the establishment of the western terminus of the Canadian Pacific Railway, and its early growth to its role as a deep-water port for commodity exports, so it is appropriate to start a review of core-periphery linkages with a description of Vancouver's strategic transportation roles.

Despite the growing significance of road transport, facilitated greatly by the strategy of successive governments to "open up" the provincial periphery via highway construction, the railways still represent crucial elements of linkages between Vancouver, the rest of the province, and indeed western Canada as a whole. Vast quantities of minerals, forest products, grains and other agricultural commodities are conveyed to Vancouver by B.C. Rail, the Canadian Pacific Railway and the Canadian National Railway.

This strategic resource export role is reflected in the operation of the Port of Vancouver (and the Fraser and North Fraser ports), Canada's largest and most important seaport by some margin. Although container traffic continues to experience sustained growth, the Port's central mission is the export of tens of millions of tons of bulk commodities to global markets. Typically export tonnage exceeds that for imports by a ratio of about 10:1 (Table 13.1). (The disparity between exports and imports in terms of value is however much less, owing to the much higher proportion of end-products and value-added goods among imports.)

Table 13.1   Inward and outward bound cargo traffic for the
                   Port of Vancouver, 1980-1993 (metric tons)

Year	Inward	Outward	Total
1993	6,510,000	54,252,000	60,762,000
1992	5,817,000	57,489,000	63,306,000
1991	6,104,000	64,610,000	70,714,000
1990	6,511,000	59,937,000	66,448,000
1989	6,351,000	57,674,000	64,025,000
1988	6,196,000	65,120,000	71,316,000
1987	5,806,000	58,151,000	63,957,000
1986	5,436,000	52,157,000	57,593,000
1985	5,338,000	50,765,000	56,103,000
1984	6,018,000	53,279,000	59,297,000
1983	4,641,000	47,007,000	51,648,000
1982	4,596,000	44,484,000	49,080,000
1981	5,138,000	44,357,000	49,495,000
1980	5,478,000	43,726,000	49,204,000

Source: Vancouver Port Corporation: City of Vancouver Economic Development Office.

The export of resource commodities from the Port of Vancouver will certainly remain a key element of core-periphery linkages well into the next century, but it may be possible to discern potentially significant changes in the nature of this relationship. Whether or not the decline in export tonnages (Table 13.1) represents the start of a trend or merely another cyclical downturn, it is clear that lower resource harvesting allowances in resource sectors such as forestry and fisheries will impose a more secular downward pressure on resource exports. Competitive pressures may also impact the Port's commodity export role, as witness the recent diversion of potash shipments from Vancouver to Puget Sound ports.

In the case of the City of Vancouver, a reshaping of the central waterfront, increasingly favouring cruise ship facilities, container traffic, residential and recreational uses, may presage a reconfiguration of Vancouver's historical transportation roles. The number of cruise ship passengers (using the Vancouver central waterfront facilities almost exclusively) has increased from under 75 thousand in 1980, to over half a million in 1993 (City of Vancouver, 1994).

Perhaps the most consequential redefinition of Vancouver's strategic transportation roles, however, lies in the expansion of Vancouver International Airport. While the Port of Vancouver's trade has remained relatively stable over the past decade, air passenger traffic has increased steadily over the same period (Table 13.2). To be sure, Vancouver International Airport's role entails an important network of air services between Vancouver and other points within British Columbia, adding another dimension to core-periphery linkages within the province, but international and, more specifically, Pacific Rim traffic is the propulsive feature of Vancouver International Airport's growth and development. Indeed, Vancouver International Airport hosts more Asian-Pacific carriers than does Sea-Tac Airport in Washington State, a facility with substantially more total traffic. Here, Vancouver International Airport is seen as a key point of connectivity between Vancouver and distant markets, cities, and cultures, reflecting Vancouver's ongoing external reorientation. Underscoring this point, a City of Vancouver staff report recommending Council support for Vancouver International Airport's second parallel runway suggested quite explicitly that the Airport would be as crucial to Vancouver's development in the next century as the Port has been in the present one (City of Vancouver, 1991a).

Table 13.2   Growth in numbers of scheduled airline passengers for Vancouver International Airport, 1980-1993

Year	Scheduled Airline Passengers	Year	Scheduled Airline Passengers	Year	Scheduled Airline Passengers
1993	9,912,429	1986	8,385,358	1983	6,370,000
1992	9,268,147	1987	8,023,520	1982	5,858,000
1991	9,380,000	1986	8,385,358	1981	6,818,000
1990	9,912,429	1985	7,005,802	1980	6,777,000
1989	9,651,237	1984	6,895,000		

Source: City of Vancouver Economic Development Office.

Production Linkages: Resource Inputs to the Industrial Metropole

Core regions within the Innisian staple model function *inter alia* as industrial metropoles, transforming natural capital extracted from hinterland regions into economic capital via secondary processing and manufacturing activities. This designation is certainly applicable in the present case, given that a succession of large-scale resource processing and manufacturing industries have figured prominently in Vancouver's economic development over the past century and, indeed, continue to have resonance even for the city's contemporary, substantially tertiarized, economy.

Many of these industries were directly linked to the provincial staple sector and were dependent upon the supply of resources imported not just from proximate areas, but also from the far reaches of British Columbia's peripheral hinterland. These industries included fish canneries, sawmills, tanneries, and agricultural processing operations situated on Burrard Inlet and in False Creek and along the Fraser River.

The immediate postwar period saw a substantial expansion of industrial production associated with British Columbia's staple sector, especially in forestry and wood products. With respect to the industrial geography of this activity, many of metropolitan Vancouver's largest sawmills and plywood and paper plants were located in a strip of industries along the North Arm of the Fraser River from southeast Burnaby to New Westminster and Fraser mills, while another dozen lumber companies were situated on the Surrey side of the Fraser (North and Hardwick, 1992: 202-203). The Vancouver local of the International Woodworkers of America (IWA) emerged as the largest in the continent (Hutton, 1985) reflecting the unrivalled scale of wood industries in Greater Vancouver and, more specifically, the crucial production linkages between Vancouver and the resource-rich hinterland.

Contemporary profiles of Vancouver's economy tend to emphasize the ascendancy of specialized service industries and occupations and, to a lesser extent, advanced-technology manufacturing, but the present-day importance of resource processing and manufacturing should not by any means be discounted. The North Fraser and suburban municipalities continue to be home to large concentrations of forest and wood products industry. But even in the City of Vancouver, where the process of economic tertiarization is most advanced, and where both market forces and public policy (in the form of rezoning and land use changes) have impelled the contraction of heavy industries and attrition within the industrial land inventory, important pockets of resource processing and production remain: six of the region's largest sawmills, for example, can be found along the North Arm of the Fraser River, and within the City's boundaries (City of Vancouver, 1993).

The residual presence of traditional resource processing and manufacturing notwithstanding, we can discern structural changes in Vancouver's production sector that imply a diminution of its linkages with the provincial staple base. These changes include, at the broadest level, a continuing process of economic restructuring and division of labour favouring service industries and occupations, a shift from heavy industry to a more diversified range of production industries within the manufacturing sector and, finally, a progression from "propulsive" industries heavily reliant upon raw resource inputs to value-added production activities deploying higher ratios of capital, technological and skilled labour inputs. In the aggregate, these trends tend to reposition Vancouver from a centre specializing in resource processing to a more diversified (but still relatively buoyant) manufacturing region, including the emergence of new production activities and zones both in the City of Vancouver and in suburban districts.

The broad dimensions of metropolitan Vancouver's changing division of labour are presented both in terms of major industry groups (Table 13.3) and principal occupational categories (Table 13.4). Table 13.3 shows the rapid growth of service industry employment over the past three decades. As a particularly dramatic demonstration of the dynamism of the advanced services especially, we can observe that the business services labour force (roughly representing "producer" services) now approximately equals the entire manufacturing labour force of metropolitan Vancouver, whereas in 1961 it was only about one-seventh as large.

Table 13.3   Labour force growth and change in Metropolitan Vancouver by industry: 1961, 1971, 1981, 1986, and 1992

	Labour Force (in 000s)					1986/92	1961/92
Industry Group	1961	1971	1981	1986	1992	*% Change*	*% Change*
Agriculture	3.8	5.4	7.7	10.2	8.9	-12.7	134.2
Forestry	2.5	3.4	4.2	3.9	3.3	-15.4	32.0
Fishing & Trapping	1.8	1.6	2.3	2.7	3.2	18.5	77.7
Manufacturing	57.5	78.7	96.3	91.1	76.1	-16.5	32.3
Construction	19.9	32.0	44.6	46.1	64.5	39.9	224.1
Transportation & Communication	34.9	49.9	69.6	73.0	73.7	0.9	111.2
Trade	59.9	85.7	126.5	136.8	187.6	37.1	213.2
Finance, Insurance & Real Estate	15.9	28.1	47.3	53.6	71.7	33.8	350.9
Services to Business	7.9	18.2	n/a	49.9	72.6	45.5	818.9
Education & Related Services	12.1	27.5	41.0	41.9	52.9	26.3	337.2
Health & Welfare Services	12.8	30.3	52.6	60.9	78.4	28.7	512.5
Accommodation & Food Services	12.9	22.3	42.6	58.7	67.8	15.5	425.6
Public Administration & Defence	18.0	22.3	37.9	39.4	38.3	-2.8	112.7

Sources: Census of Canada, 1961, 1971, 1981; 1986 mid-term census; 1992 estimates prepared by Labour Force Survey Subdivision, Household Surveys Division, Statistics Canada.

Occupational data offer another measure of the distinct nature of tertiarization in the Vancouver case. Table 13.4 shows that employment in the key "managerial, administrative and related" category has exhibited sustained growth even during the 1981-86 period when British Columbia experienced the most serious recession since the 1930s.

Table 13.4   Employment change in
                  Vancouver CMA by occupation (1971, 1981, 1986)

Occupational Category	Employment			Change (%)	
	1971	1981	1986	1971-81	1981-86
Managerial, administrative and related	21,030	52,805	85,020	151.1	61.0
Natural sciences, engineering, mathematics	14,215	25,190	26,220	77.2	4.0
Social sciences and related fields	5,440	12,720	17,530	133.8	37.8
Occupations in religion	795	1,040	1,455	30.8	39.9
Teaching and related occupations	16,135	25,260	25,915	56.6	2.6
Medical and health	19,495	32,920	37,960	68.9	15.3
Artistic, literary, recreational	5,315	10,370	14,770	95.1	42.4
Clerical and related occupations	88,700	143,290	146,780	61.5	2.4
Sales occupations	55,855	75,330	81,390	34.9	8.0
Service occupations	58,715	85,355	108,120	45.4	27.1
Farming, horticultural	7,760	9,600	12,295	23.7	28.1
Fishing, hunting, trapping	1,795	1,295	2,155	-27.9	66.4
Forestry and logging occupations	2,390	2,270	2,320	-5.0	2.2
Mining and quarry (including oil and gas)	1,130	935	790	-17.3	-16.5
Processing occupations	18,365	21,410	19,650	16.5	-16.6
Machining and related occupations	11,794	14,090	11,640	19.5	-17.4
Product fabricating, assembling	28,935	40,785	40,690	41.0	-0.3
Construction trades	32,140	41,705	40,980	30.0	-1.7
Transport equipment operating	19,170	25,285	27,510	31.9	8.8
Material handling and related not elsewhere classified	15,865	17,340	15,410	9.3	-11.5
Other crafts and equipment operating	5,730	7,730	5,925	28.0	-19.2
	430,769	**646,725**	**724,525**	**50.1**	**12.0**

Source: Census of Canada (1971, 1981, 1986).

Reference to the industrial profile of companies also tends to underline the importance of services for Vancouver. Unlike other cities within which leading industrial concerns constitute propulsive firms for the regional economic base, Vancouver's largest firms tend to be service entities, both in the public and private sector. A recent survey of the 38 largest employers in Metropolitan Vancouver (that is, with over 1,000 employees) included only one resource-manufacturing corporation, MacMillan Bloedel Ltd., in 30th place, just behind the Hong Kong Bank of Canada (City of Vancouver, 1994).

As noted, shifts within Vancouver's manufacturing sector also suggest implications for core-periphery production linkages in British Columbia. Here, we can discern a shift from a concentration in heavy manufacturing to light industries, and (in some respects) to a more diversified production base. In 1956, three industries—wood products, metals manufacturing, and food and beverages—comprised 41 percent of Vancouver's manufacturing employment. By 1991, a four-industry group consisting of printing and publishing, food and beverages, garment production, and wood products accounted for 67 percent of Vancouver's manufacturing employment (City of Vancouver, 1993). Employment in the wood products sector declined from 17,800 in 1956 to 12,200 in 1991, attributable both to relocations and outright closures and also to more intensive applications of capital during the last decade. Vancouver's industrial development in recent decades also implies less reliance on proximate access to the province's staple resource base. Table 13.5 shows considerable concentration in the four industrial categories cited above, but with representation across a quite varied range of SIC groups.

At another level, the emergence of "new production spaces" (after Scott, 1988) reveals a maturing and diversifying Vancouver industrial economy with diminished requirements for massive raw resource inputs characteristic of Vancouver's heavy manufacturing phase in the immediate postwar period. These include, for example, the emergence of over one hundred "applied design firms" which impose a new production regime in the old inner city industrial districts of the City of Vancouver (Hutton, 1994a), electronics and telecommunications firms within the inner suburban precincts of Richmond and Burnaby especially, and the rapid growth of garment production in East Vancouver (Figure 13.1). Certainly, each of these dynamic industries requires material inputs but, arguably, access to skilled labour, technology, and offshore-sourced inputs are crucial to each.

Consumption Linkages:
Staples and the Metropolitan Consumer Market

As observed above, staples extracted within British Columbia's periphery have been important stimuli to industrial production in Vancouver over the past century. In framing the dimensions of the core-periphery model, however,

Table 13.5   Manufacturing establishments in
                   Metropolitan Vancouver, by SIC (1994)

SIC	Manufacturing Group	Listings	Percent
2000-2099	Food and kindred products	399	5.9
2100-2199	Tobacco	1	0.0
2200-2299	Textile mill products	76	1.1
2300-2399	Apparel and other textile products	336	4.9
2400-2499	Lumber and wood products	700	10.3
2500-2599	Furniture and fixtures	245	3.6
2600-2699	Paper and allied products	116	1.7
2700-2799	Printing and publishing	1,093	16.1
2800-2899	Chemical and allied products	257	3.8
2900-2999	Petroleum and coal products	21	0.3
3000-3099	Rubber and plastic products	184	2.7
3100-3199	Leather and leather products	33	0.5
3200-3299	Sand, clay and glass products	248	3.7
3300-3399	Primary metal industries	119	1.8
3400-3499	Fabricated metal products	565	8.3
3500-3599	Machinery, except electrical	523	7.7
3600-3699	Electrical and electronic equipment	288	4.2
3700-3799	Transportation equipment	849	12.5
3800-3899	Instruments and related products	145	2.1
3900-3999	Miscellaneous manufacturing	592	8.7
	Total	**6,790**	**100.0**

Source: Contacts Target Marketing: City of Vancouver Economic Development Office.

it is important to acknowledge the role of staples in consumption (final demand), as well as representing crucial inputs to industrial production.

This interpretation acknowledges the vital role of the provincial hinterland in supplying foodstuffs, water, energy and other commodities to the population of a rapidly-expanding metropolitan core region. Vancouver's capacity to provide for its own consumption needs from local sources is constrained not only by physical and political geography in the form of mountains, the sea, and the international border immediately adjacent to the suburban municipalities of Surrey and Delta, but also by the relentless incursion of urban land uses into the agricultural areas of the Lower Fraser Valley. Thus, Vancouver must import vast amounts of food and other consumption staples from outside the region, including, for example, a considerable quantity of vegetables and fruit from the Okanagan Valley and dairy products from Vancouver Island. In empirical

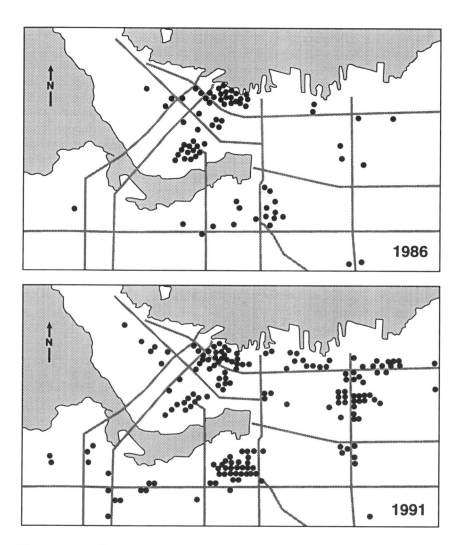

Figure 13.1   Spatial pattern of garment firm expansion in
Vancouver's metropolitan core, 1986-1991

terms, a recent study of the "appropriated" carrying capacity of urban areas
determined that the 1.7 million people of metropolitan Vancouver require an
area 18 times larger than the Lower Fraser Valley for food, forest products, and
energy (Wackernagel, 1994).

While Vancouver's consumption interdependencies are an important dimension of core-periphery relationships, again we can discern some significant
(if in some respects incipient) aspects of change here. With respect to the crucial food supply question, Vancouver is now more heavily reliant upon foreign

sources, notably, fruit and vegetables from the Central Valley of California, Mexico, and even further afield.

One final important consumption linkage is in the form of the purchase of recreational property by Vancouver residents in places like Vancouver Island, the Gulf Islands, Sechelt, and the south Okanagan, which suggests a further refinement and redefinition of relationships between Vancouver and the rest of British Columbia. Vancouver's association with the rest of the province can thus experience a significant measure of deepening and transformation as well as disengagement.

Sociocultural Linkages: The Staple Sector and Urban Class Formation

While core-periphery linkages are conventionally interpreted in terms of commodity flows, production linkages and control functions, there are also significant social and cultural implications. In the case of British Columbia, the development of a massive staple economy within the provincial hinterland influenced the formation of specific class and community structures within metropolitan Vancouver. Within Vancouver's production sector, staple processing has created occupational divisions such as plant managers, technicians, secretarial staff and, of course, operatives, while in corporate head offices, important social groups include board members and executives, cadres of middle managers and supervisors, and large numbers of clerical workers. Many other employees in public utilities and crown corporations (such as B.C. Hydro) situated within Vancouver are also highly dependent on British Columbia's resource sector. At opposite ends of the socioeconomic hierarchy within the broadly-defined resource community are, at the top end, owners of capital, and, at the bottom, (mainly older) workers displaced by technology or by injury. From these numerous subdivisions, we will focus upon three principal expressions of sociocultural relationships within the context of core and periphery in British Columbia: the cadre of senior management of Vancouver-based resource companies, the resource-processing and manufacturing workforce resident in Vancouver, and a population cohort resident in the Downtown Eastside district consisting, largely, of former staple industry workers.

These three Vancouver social groups, each closely associated with the provincial resource sector, represent distinctive tiers in the city's socioeconomic hierarchy and the relative positioning of classes within a capitalist and corporate urban culture. That said, recently there have been changes within this hierarchy reflecting larger transformations within the economy of the province and, more specifically, the fortunes of British Columbia's staple sector.

First, at the peak of Vancouver's social hierarchy, the growth of large, integrated resource companies in British Columbia has given rise both to the owners of capital and, also, to a much larger corporate elite consisting of directors and senior officers. These (overwhelmingly male) executives tended to occupy the corner offices of the most prestigious, "Class A" high-rise office buildings in the heart of Vancouver's Central Business District. From the bastions of the corporate complex, the affairs of British Columbia's resource giants were, to a large extent, determined by this exalted stratum of senior decision-makers. It must of course be acknowledged that this power of "command and control" over the province's staple sector was necessarily shared with external agencies and actors. As Vancouver functioned as core to the peripheral hinterland, British Columbia as a whole represented a marginal element of the Canadian (and increasingly continental) industrial economy: head offices in the Canadian heartland centres of Montréal and Toronto exercised considerable influence, while American corporations such as Crown Zellerbach were in evidence by the 1950s.

The corporate elite of Vancouver-based resource companies, in addition to representing a crucial component of the City's corporate complex, exhibited well-defined residential preferences and thus helped to define Vancouver's social geography. Aside from the relatively small cadre of company owners that inhabited the most exclusive districts of Shaughnessy (with additional enclaves in West Point Grey and West Vancouver), many of the senior officers and board members resided in the highly desirable Kerrisdale and Point Grey neighbourhoods. A related class of business people engaged in the financing of resource development in British Columbia, including stockbrokers and members of the Vancouver Stock Exchange, displayed somewhat similar residential locational preferences, but with a stronger inclination for the British Properties district of West Vancouver. Known popularly among long-term residents of Vancouver as "martini slopes," the British Properties, with its profusion of swimming pools and other conspicuous manifestations of recently-acquired wealth, was thought to represent a distinctly *parvenu* and even somewhat raffish contingent of speculators and stock promoters, distinct from the more established resource corporate elite.

Certainly this class (or classes) of individuals associated with the control of British Columbia resource companies still retain considerable wealth, prestige, and influence. During the last decade, however, there has been substantial pressure on this privileged socioeconomic elite because of both cyclical downturns and the sequence of corporate mergers and acquisitions that accelerated during the 1980s, which was part of a broader restructuring of resource companies.

If the last 15 years has seen at least a relative diminution in the position of resource company executives within the upper echelons of Vancouver's corporate elite, the impacts of change on less powerful elements of Vancouver's "staple sector culture" have been even more severe. The redeployment of production capital and, more specifically, the relentless introduction of new technology in the resource manufacturing sector, have drastically shrunk Vancouver's wood industry workforce. This trend is especially advanced in the

City of Vancouver which contained a massive industrial plant of wood processing and manufacturing operations both in the urban core and along the North Arm of the Fraser, but which now has only about one-fifth of the metropolitan area's forest industry workforce, that is, considerably less than the City's current share of Vancouver CMA's manufacturing labour force (City of Vancouver, 1993). The contraction of Vancouver's forest products labour force is also reflected in the residential landscape: many of the old blue-collar districts of the east side present the signs of (at least incipient) gentrification, while in southeast Vancouver, close to the mills of the Fraser River, the former concentrations of IWA workers and their families have given way to a dramatic influx of new immigrants, especially ethnic Chinese and South Asians. Some of the latter still work in the mills, to be sure, but overall the sociocultural transformation of southeast Vancouver ranks among the most rapid in the contemporary experience of urbanization in Canada, and offers yet another aspect of change in Vancouver's broader relationship with the province's staple economy.

A third and final example of sociocultural linkages between Vancouver and the provincial periphery is found in the Downtown Eastside, the most vulnerable community in Vancouver and, by some accounts, possibly the poorest neighbourhood in Canada (Ley, 1992). Although distant from the palatial homes and elegant lifestyles of the corporate elite in exclusive Shaughnessy and Kerrisdale, many of the residents of the Downtown Eastside share with these privileged groups a close working experience with British Columbia's staple economy. Until recently, certainly, the typical resident of the Downtown Eastside was a middle-aged or elderly male with a record of employment in one or more resource industries, as a miner, forestry worker or commercial fisherman (Hulchanski, 1989: 3).

While not an affluent precinct by any Canadian standards, the Downtown Eastside has nonetheless provided a long-term and relatively stable neighbourhood for many of these former resource sector workers over the last several decades. Despite the stereotyped image of transience, many of the residents of this community have achieved some measure of long-term, if necessarily tenuous, residency. While incomes are low (and for many variable), there is a comparatively rich assortment of 40 or so social service agencies present, or active in, the Downtown Eastside (City of Vancouver, 1986). These include not only governmental agencies and NGOs, but also the Downtown Eastside Residents Association (DERA), one of Canada's most durable and successful inner city social service and advocacy groups. DERA has managed to secure a substantial amount of resources for the residents of the Downtown Eastside, including affordable housing, has effectively voiced the area's concerns at City Council and in other arenas, and has generally served to organize and "anchor" the residents of this community.

The Downtown Eastside represents one element of sociocultural relationships between Vancouver and the hinterland staple economy but, again, we can detect the beginnings of important changes. First, the composition of the community itself is changing, both as a consequence of the aging of long-term cohorts resident in the area, and also the recent introduction of new groups, including younger people, couples and even families. DERA has also expressed concerns that new market housing development proposed for the Downtown Eastside will potentially displace existing residents and destabilize this vulnerable community (*Vancouver Sun*, 14 II 1995: 1312). Indeed, gentrification is already occurring here, presaged by an influx of artists and loft-studio conversions from the early 1980s.

In addition to these *in situ* trends, the Downtown Eastside continues to experience pressure from large-scale redevelopment in adjacent districts. This pressure has accelerated over the last decade, since the Exposition of 1986 which saw evictions of long-term residents from a number of single-room occupant residential hotels (SROs). Currently, the Downtown Eastside is surrounded by such projects as Concord Pacific's Pacific Place, the "CityGate" project (which includes a social housing component), and the ongoing redevelopment of the Central Waterfront, as the comprehensive reconstruction of Vancouver's core proceeds apace. The City of Vancouver has tried to stabilize the area and enhance security of tenure for long-term residents of the Downtown Eastside (and nearby Victory Square), by supporting social housing and by limiting redevelopment potential. There is real concern about the prospects for the residents of this marginalized precinct of the City, especially in light of the declining senior government commitment to social housing, and the scale of proximate redevelopment projects. Spillover effects from the latter are likely to include land price and rent inflation and speculative private investments. Combined with the natural aging process already cited, these trends will inevitably reshape the social profile of the Downtown Eastside, and the inner city more generally, and in all probability will lead to a relative or even absolute diminution of the number of former resource industry workers that have in the past represented an important link between core and periphery within a staple economic system.

Command and Control: Specialized Service Linkages

Each of the four key dimensions of connectivity addressed above (transportation, production, consumption, and social structure) are important, and are indeed essential to a rounded profile of interdependency and linkage, but specialized services (head office and senior management, finance, and business or "producer" services) represent the most crucial attributes of core-periphery relationships. While each of the four aspects of linkage previously discussed emphasize two-way commodity flows between the resource hinterland and the metropolitan area, specialized service flows (directives, market intelligence, information and knowledge) are overwhelmingly transmitted from the core to the periphery. Further, these higher-order service functions are critical to an appreciation of staple economic systems, because they:

(a) embody the "command and control" functions by which the metropole directs the fortunes of the staple sector, with respect to the deployment of capital and labour, marketing decisions, and other critical operations;

(b) represent the most crucial aspect of asymmetry between core and periphery within staple economic systems, reflected in divergent growth rates, cyclical stability, and other measures;

(c) typify the more diversified and advanced (tertiarized) economic base of core regions vis-à-vis the more truncated industrial structure of hinterland regions;

(d) imply imbalance within the inter-regional "terms of trade," that is, the higher value-added character of specialized services, vis-à-vis raw commodity shipments;

(e) reflect the leakage of development benefits from the hinterland to the metropolitan core region.

In short, specialized services (administration, control, information, technology) represent not merely one of many inter-regional linkages and flows, but, rather, a defining feature of the core-periphery structure in a staple economy setting.

While the downtown typically contains a range of economic activities, including business, retail and personal services, and even in some cases speciality manufacturing (such as jewellery and fashions), the crucial component is an intensely interactive agglomeration of head office and senior management operations, banking and finance (and, more specifically, commercial, merchant and international banking), and producer or business service activities. This tripartite ensemble of head offices, financial institutions and producer services is easily discernible in the City of Vancouver, and linkages to the provincial staple sector and peripheral regions are manifestly of great importance. A study published about a decade ago disclosed that the downtown peninsula of the City "... contains the head offices of virtually all the province's major business corporations; over half of them are directly implicated in the resource economy, particularly as lumber and mineral producers" (Ley and Hutton, 1987: 417). Vancouver's downtown corporate complex also contains financial institutions closely associated with the resource industry, most notably the Vancouver Stock Exchange (VSE), as well as brokerage houses and commercial finance divisions of the major national banks.

By far the most numerous elements of the corporate complex, however, are producer services such as corporate legal and accountancy firms, management consultants, and consulting engineers. There are many hundreds of such enterprises within the corporate complex and, indeed, the rapid growth of producer services represents a propulsive influence on the expansion of office space in the downtown over the past two decades (Hutton and Ley, 1987). Certainly, the large resource company headquarters have been seen as the flagships of the CBD and have for much of the postwar period defined the character of the downtown. As North and Hardwick observe, resource corporations "shaped the landscape" of the central city by virtue of the construction of purpose-built office buildings, including the MacMillan Bloedel building on Pender Street, B.C. Forest Products on Melville Street, Rayonier on Georgia Street, and the B.C. Electric Company on Burrard (North and Hardwick, 1992: 206). Later, MacMillan Bloedel helped to redefine Georgia Street, the Central City's principal east-west thoroughfare, by constructing a distinctive high-rise tower, projecting its dominance both in the city and in the more peripheral areas of the province.

The operations and pattern of business services in a regional setting are depicted in the (in some ways analogous) case of Montréal in Figure 13.2. This model illustrates the direction and relative magnitude of business service flows, with Montréal clearly the command and control centre of its regional hinterland, although there is some penetration of the Montréal-centred region by the larger business centres of New York and Toronto, primate cities within the national economies and urban systems of the U.S. and Canada, respectively. This, in turn, introduces the notion of hierarchy among cities within national (and increasingly international) levels distorting or modifying the containment of inter-regional linkages within core-periphery structures. The most highly-specialized firms (in terms of expertise and knowledge) can clearly compete against even the largest indigenous centres within the latter's hinterland, as evidenced in the increasing internationalization of specialized service provision (Daniels, 1993).

Such a modification of metropolitan-hinterland linkages is present in British Columbia, where the parameters of core-periphery relationships have been subject to the external influence of more dominant metropoles. Indeed, ". . . because control of many of B.C.'s major firms lay outside the province, [Vancouver] played only an intermediary role and was itself greatly affected by decisions made in other metropoles" (Barnes et al., 1992: 181-2). If anything, the integrity of Vancouver's command and control functions within the classic core-periphery structure are under increasing pressure from external factors.

The galaxy of influences are best presented according to the spatial scales at which they operate, although some care must be taken to acknowledge the complexities of interdependence and connectivities among scales. First, there are the changes that pertain to large corporate enterprises, including resource companies at the level of economic globalization and market integration; a second set of factors relate to the level of the province, particularly the fortunes of British Columbia's staple sector in light of serious resource depletion

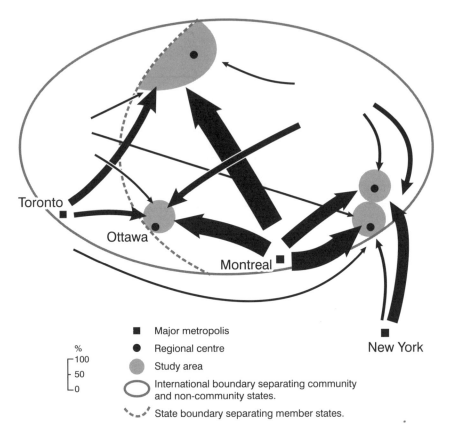

Figure 13.2   Business service flows between sub-regions of Quebec
(after Polèse, 1981)

and changing social attitudes toward environmental enhancement; while a third
and final set of forces concerns processes of restructuring and reorientation
within Vancouver, the provincial "core," itself. Cumulatively, these influences
must reshape the nature of metropolitan-hinterland relationships in a region
so manifestly open to exogenous change and, more specifically, the control
functions which are seen to represent a defining attribute of core-periphery
relations within staple economies.

Macro-level shifts: Globalization, integration and the
spatial reorganization of corporate control

The preceding discussion established that, rather than exercising complete
dominance over a captive and largely contained hinterland, core regions must
in many cases be reconciled to the intervening power of even larger control
centres. Over the last decade and a half, processes of deregulation, market

integration and corporate restructuring have greatly expanded arenas of command and control, characterized in part by mergers, acquisitions and relocations, and the reconsolidation of decision-making within global cities (Sassen, 1994). In the Canadian context, major provincial centres, such as Montréal and Vancouver, have lost considerable ground to Toronto, indisputably the primate business and financial centre of Canada (Coffey, 1994).

Corporate restructuring is by no means new in the resource sector; indeed, MacMillan Bloedel, British Columbia's largest forest industry company, is itself the product of a merger (in 1965). But a series of acquisitions and relocations over the 1980s and the present decade have served to vitiate Vancouver's role as control centre for the provincial staple economy. The previous decade saw, for example, the acquisition of B.C. Forest Products by Fletcher Challenge, a multinational conglomerate with control centres in New Zealand and San Francisco, emphasizing the international dimension of restructuring. More recently, in April of 1995 Scott Paper (which has itself been the subject of a buy-out) announced its intention to relocate its Canadian head office (and over 80 employees) from Vancouver to Toronto, closer to its production facilities and largest markets, again underscoring Toronto's primacy within the national arena, even in the aftermath of the severe downturn that afflicted much of Ontario in the early 1990s.

Globalization and market integration also carry with them alterations in the dynamic of supply and demand in the resource sector that will continue to impact the fortunes of staple production in British Columbia. With respect to the forest products and wood industries sector, the emergence of low-cost competitors in the softwood lumber business (e.g., in the southeast U.S.) will have implications for British Columbia, and the looming presence of the Russian Far East, with its massive reserves of timber and other staples, must also be taken into consideration. Within the mining sector, a redirection of exploration capital from British Columbia to Chile and other South American countries, coupled with increasing restrictions on mining activity in British Columbia, resulted in a one-sixth reduction in mining employment between 1990 and 1992 (Kunin and Knauf, 1992), and is likely to result in contractions within Vancouver's corporate complex.

Indeed, there is empirical evidence to point to reductions both in mining and forestry office staff and, as a result, Vancouver's downtown has an increasingly truncated resource sector head office organization. A recent telephone survey conducted by the author, including approximately 20 large resource firms, revealed long-term as well as more recent downsizing of both forestry and mining companies within Vancouver's corporate complex. Additional corroborative evidence concerning the impacts of corporate restructuring was generated in a recent survey of Vancouver's forest industry labour force, including head office staff as well as production workers. The study concluded that the ". . . trend toward mergers has eliminated duplication of administra-

tion and managerial functions, reducing the number of downtown office employees substantially" (City of Vancouver, 1993: 73). The total number of forest
industry head office employees in Vancouver's corporate complex was estimated to be about 1,200 in the same study—a far cry indeed from the 1970s,
during which one company alone, MacMillan Bloedel, employed the better
part of a thousand office staff in its prestigious, purpose-built tower. (MacMillan
Bloedel subsequently elected to divest itself of its Georgia Street tower and
now occupies more modestly-scaled premises, an event itself symbolic of the
changing order in the CBD.) Clearly, these 1,200 or so forest-sector office staff
include many senior corporate executives who still exert substantial decision-
making power but now constitute only about 1 percent of downtown Vancouver's 100,000 or so office workers (City of Vancouver, 1991b). To some extent,
this downsizing reflects the more pervasive corporate restructuring process
that has seen contractions in (especially middle-management and clerical) staff,
but there can be little question that the data reflect long-term trends and at
least a relative diminution of Vancouver's traditional control functions.

Dynamics of change within British Columbia

It seems apparent that factors operating at the macro-level are impacting
Vancouver's traditional control functions, but we can also reference significant trends and events within British Columbia that will increasingly reshape
the configuration of core-periphery relations within a staple economy setting.

These trends seem certain to include an absolute contraction of primary
industries in British Columbia that may, indirectly at least, reduce the scale of
head office operations within Vancouver. A substantial shrinking of the extractive sector seems inevitable in the face of serious resource depletion (both
renewable and non-renewable), the creation of wilderness reserves within which
extraction will be prohibited, or at the very least severely restricted, and sharply
reduced allowable yields in many areas outside these designated parks and
zones of preservation.

Depletion of forest fibre has accelerated over the postwar period, and the
levels of annual harvest have roughly tripled over the last three decades (Binkley,
1992). Public concerns about the unsustainable nature of contemporary industrial forestry, coupled with overall environmental degradation associated with
mining and other staple industries, has shifted the balance toward greater
conservation. Recent policy initiatives include a provincial government commitment to expand wilderness preservation areas from 6 to 12 percent of the
province's land base, the introduction of a new Forestry Practices Code, and
the establishment of the Commission on Resources and the Environment (CORE)
as an adjudication tribunal mandated to balance the goals of stakeholders and
to produce regional land use plans within British Columbia. Continuing land-
claim negotiations and treaty processes between the Government and the First
Nations groups have a strong potential of removing a considerable portion of

territory from the land base traditionally accessed by forest and mining companies, although native groups are themselves likely to undertake their own resource industry developments. In these cases, however, new forms of First Nations tenure over extraction will also likely have the effect of diluting the control power of Vancouver-based firms over the hinterland.

The next decade will also see a new regional and community-level economic geography within British Columbia that will again redefine the nature of core-periphery relations and, more specifically, the linkages of specialized services from the metropolis to hinterland centres and regions. Although metropolitan Vancouver has been steadily increasing its share of the province's total population and employment over the past decade (Davis and Hutton, 1992), this period has also seen substantial economic expansion in regions within southern and southwest British Columbia (notably the Okanagan Valley, the Sechelt Peninsula, and southern and eastern Vancouver Island) as well as the growth and development of several cities with a population of 60,000 or greater outside the Lower Mainland. With only one-thirtieth of Vancouver's population, these centres cannot, of course, compete with the metropolitan core in terms of the most specialized service and financial activities, but they will attract more tertiary industry, raising the possibility of service import-substitution. A recent study disclosed that cities such as Kelowna, Kamloops, Prince George and Nanaimo ". . . are expanding their roles as sub-regional service centres" (Kunin and Knauf, 1992) and so core-periphery relationships may be replicated at a smaller, sub-provincial scale.

Changes to core-periphery control functions within British Columbia also include banking and financial industries. Vancouver ranks as the dominant financial centre within the province but, again, we can discern significant changes that require a reappraisal of the classic core-periphery framework.

For much of the 1970s and 1980s the principal indigenous financial institutions in British Columbia were the Bank of B.C. and the Vancouver Stock Exchange (VSE), both ensconced within the confines of Vancouver's corporate complex. Each of these financial agencies had powerful links to the province's staple economy: the VSE, despite its variable reputation, was ranked as the largest venture capital institution for resource exploration in North America, while the Bank of B.C. was established, in large part, to facilitate development in British Columbia and, in some part, to reduce the province's historical dependence on the major chartered banks domiciled in Central Canada.

Although there is inadequate scope here to present a comprehensive dissection of the experience of the VSE and the Bank of B.C., we can at least recall the basic events and shifts, together with their implications for Vancouver's role as financial centre within the specific provincial context. In the case of the former, the role of the VSE has been declining in relative and, perhaps, even absolute terms over the past decade or so. Trends toward financial consolidation and re-concentration in Canada over the 1980s greatly favouring Toronto substantially undermined Vancouver's commercial banking role, including stock exchange, brokerage, and commodities. Some of the blue-chip resource companies were listed principally on the Toronto Stock Exchange, leaving Vancouver with a residual of "junior" and, in most cases, highly speculative resource stocks (Ley and Hutton, 1991). In absolute terms, the VSE has lost about one-quarter of its total listings since 1986 (City of Vancouver, 1994).

The example of the Bank of B.C. may be even more instructive. The Bank of B.C., like at least two Alberta-based banks, was not able to survive the vicissitudes of resource industries and the limitations of regional markets, and it was acquired by the Hong Kong Bank of Canada in 1985. Although the failure of the Bank of B.C. experiment can be seen as a blow to the financial sovereignty of the province, the Hong Kong Bank of Canada (HKBC) was (and is) headquartered in Vancouver and offers the city a high profile within the national banking realm, especially within the dynamic "middle-market" sector that it sees as its principal client base. HKBC is also a major outpost of the Hong Kong and Shanghai Banking Corporation (HKSBC) headquartered in Hong Kong, testifying to Vancouver's deepening engagement with the dynamic Asia-Pacific region (although others see such trends as evidence of British Columbia exchanging a peripheral position within the Canadian confederation for an even more subordinate status within an emerging Pacific Basin core-periphery system; cf Resnick, 1985).

At the same time, it would be premature to correlate the variable fortunes of Vancouver's banking and financial sector with an overall weakening of the city's financial role. Within the national context, Simmons suggests that Vancouver has been "slowly overtaking" Montréal as Canada's second leading financial centre (Simmons, 1991), although the degree of Toronto's pre-eminence in this sphere ensures that the gradient between first and second position within the Canadian financial hierarchy is a precipitous one. Simmons attributes Vancouver's relative ascendancy within the national banking and financial domain in part to its role as provincial financial centre, but also attributes much of its recent growth to the rapid expansion of Vancouver's metropolitan market and to a larger and more diversified industrial base. To this we can certainly add intermediary financial roles associated with foreign investment in Vancouver's property market, development industry and production sector.

Vancouver' Market Reorientation: Service Externalization and Exports

Changes within Vancouver's producer service production and export orientation provide another measure of the City's evolving relationship with the provincial economy and, increasingly, more distant markets. It will be recalled from the Montréal example by Polèse elucidated earlier that dominance in inter-regional business service flows is a prime characteristic of core regions within a staple economy context. We can trace a distinct evolution in the destination of intermediate service outputs from Vancouver over a two decade period, informed by a review of several large-scale empirical studies conducted over that period.

Table 13.6 depicts schematically a three-stage model of Vancouver's producer service production and trade, including specific reference to several supporting studies. While these studies involve somewhat different methodologies and samples, a clear pattern emerges. First, an input-output study of metropolitan Vancouver by Davis and Goldberg in 1971 (involving firms equalling about one-quarter of Vancouver CMA's total employment) revealed that the bulk of Vancouver's service output (almost 90 percent) was divided virtually equally within two principal markets: the local, metropolitan market and, secondly, the rest of the province. This split clearly reflects both the traditional role of cities as higher-order central places within regional settings and, more particularly, Vancouver's role as provincial primate city.

In the second stage (Table 13.6), the provincial territory is still very important, but significant changes in the market orientation of Vancouver's producer services are becoming apparent. Here, two independent studies conducted by Hayter and Barnes and Ley and Hutton in the early to mid-1980s disclose increasing extra-provincial exports of producer services such as management consulting, consulting engineering, and the FIRE group (finance, insurance, and real estate). It is clear that this growth in intermediate service exports can be attributed not only to a widely-experienced trend toward the externalization of service provision among firms in many advanced regions (Daniels, 1993) but, also, to a more specific set of push-pull factors. The "push" influence here is construed as concern among some of Vancouver's producer service firms about the security and growth prospects for the traditional provincial resource market in light of the severe commodity price shocks and accompanying recession of the early 1980s. Correspondingly, "pull" factors can be seen as including the growing attraction of international markets, especially within the dynamic Pacific Rim region (Ley and Hutton, 1987).

In this case, the documented growth of Vancouver's producer service exports to Pacific markets should also be seen as part of a growing commitment to foster the city's role within the Asia-Pacific realm, a commitment which included substantial new public investments in Vancouver, such as the 1986 Exposition, rapid transit, and a downtown convention centre. A final study, conducted in 1990, confirmed the growing Pacific market orientation of Vancouver's producer services, attributable to increased government support (at all levels) for service export sales and marketing, as well as more vigorous market exploration efforts of private firms themselves (Davis and Hutton, 1991). Implicit in these trends is diminished reliance upon traditional local markets for Vancouver's producer services. This may be seen as following the broad pattern of other jurisdictions, for example, in the case of New York where legal and other business services were able to thrive despite the loss of headquarter

Table 13.6   The evolution of Vancouver's service production and trade,
c. 1970-1990

I C.1970 *Regional* *Central place* *Hinterland* *control centre*	II 1980s *'Transitional'* *Service Export Phase*	III 1990 *Pacific Rim City*
1. 'Macro-level' Trends and Processes		
Tertiarization (urban economy and labour force)	*Externalization* (specialized service inputs)	*Globalization* (markets)
2. Public Policy and Regulatory Influences		
Government investment in infrastructure in hinterland	Market deregulation; privatization Government investment in Vancouver: Expo 86, rapid transit, convention centre	Government support for export marketing
3. Key Attributes and Characteristics		
Vancouver as 'downtown B.C.' Expansion of CBD 'corporate complex'	Exploration of export markets (domestic and international) Response to concerns about provincial economy (cf commodity price shocks), attraction of Pacific Rim markets	Vancouver increasingly integrated within Pacific markets, societies, network of trading and 'Gateway' cities
4. Empirical References		

Davis & Goldberg I-O
study of metropolitan
Vancouver
economy (1971)
- 44% of services to
  local market
- 44% to ROP
[Davis, 1976]
	1. Hayter & Barnes survey:
	   observed increase in
	   service export ratios
	   [Hayter and Barnes, 1991]
	2. Ley and Hutton survey
	   (n=626 firms)
	· 10% of services to
	   Canada outside B.C.
	· 7% of services exported
	   internationally
	[Ley & Hutton, 1987]
		Davis and Hutton
		survey (n=261)
		· increasing presence in
		  Asia-Pacific markets
		· internationalization
		  of producer services
		[Davis & Hutton, 1991]

Table 13.7   The Innisian core-periphery framework revisited: Summary table of Vancouver's changing relationship with British Columbia's staple economy

	I *'Classic' Core-periphery Structure*	*Transitional Processes and Influences*	*II* *'Restructured' Core-periphery Framework*
TRANSPORTATION LINKAGES	· core region as entrepôt and point of export for bulk commodities (import of finished goods) · key facility: seaport	· globalization of resource sector · deregulation, market integration · resource depletion in 'mature, advanced' staple regions	· decentralization of bulk cargo facilities · redevelopment of central waterfront · key facilities: seaport, airport and telecommunications
PRODUCTION LINKAGES	· core region as 'industrial metropole' · emphasis on resource processing and manufacturing · reliance on hinterland for resource inputs	· tertiarization of urban economy · post-Fordism, flexible specialization · maturation/growth of metropolitan economy · resource depletion	· more diversified urban industrial base (industries, markets) · greater reliance on skilled labour, entrepreneurial, technological production inputs · 'loosening of bonds' between core and periphery
CONSUMPTION LINKAGES	· core region dependent on the periphery for consumption staples: water, foodstuffs, energy	· resource depletion, changing land use in hinterland · new extra-regional staple sources for food, and other staples	· growth of urban 'appropriated carrying capacity' for basic consumption items · diminished reliance on hinterland
SOCIO-CULTURAL INTERDEPENDENCIES	· influence of staple economy on class and community formation may include: owners of capital, managerial and professional groups, operatives, etc.	· 'staple culture' under pressure: corporate reconcentration and restructuring, higher capital ratios in the workplace, etc.	· diminution of role of certain classes · rise of entrepreneurs, 'new middle class' (Ley) in urban society · gentrified urban landscapes
SPECIALIZED SERVICE LINKAGES	· core region as 'command and control' centre for extensive staple economy · key role of 'corporate complex': HQ operations, banking and finance, producer services	· accelerating series of corporate mergers, acquisitions, relocations in 1980s · telecommunications enabling more centralized control in national primate and 'world cities' · externalization of services	· penetration of regional staple economies by MNEs · control functions vitiated in 2nd-3rd order core regions · producer services in core regions seek new export markets

firms, which represented their original markets, by securing a new client base as distant as Denver (Conservation of Human Resources, 1977). In the Canadian context, Montréal and Vancouver producer services firms have, to some extent, overcome both the limitations of provincial resource markets and the primacy of Toronto in national corporate control, by exploring international market opportunities in areas such as consulting engineering.

CONCLUSION: TOWARD A RESTRUCTURED
CORE-PERIPHERY FRAMEWORK IN BRITISH COLUMBIA

This chapter has presented an examination of five important elements of core-periphery linkages between Vancouver and the rest of British Columbia and, more specifically, the nature of contemporary change in each of those elements that portend a substantially restructured relationship (or set of relationships). Particular attention was directed to an analysis of specialized service linkages and, more specifically, the "command and control" functions which perhaps most crucially define the asymmetrical nature of the core-periphery structure in favour of the metropolitan core.

Although the empirical evidence available is somewhat threadbare or fragmentary in places, the discussion points quite clearly to comprehensive changes in the tight bonding between core and periphery in British Columbia. Some of these changes are represented in summary form in Table 13.7, with a view to capturing essential features of convergence and re-association between core and periphery in British Columbia. In each of the five identified dimensions of linkage and interdependency—production, consumption, community and social structure, and specialized services—we are presented with clear evidence of a new, and dynamic restructured core-periphery framework.

As observed earlier, however, as comprehensive and potentially profound as these changing relationships between core and periphery are, it would be premature to speak in terms of Vancouver being "decoupled" from the rest of the province. Vancouver's control functions, although weakened by the series of corporate mergers, acquisitions, relocation and restructuring of the past decade and a half, are still important. The export of British Columbia's bulk commodities is still very much the principal activity of the Port of Vancouver. In other spheres, core-periphery relationships may be renegotiated. On the consumption side, for example, the increasing presence of Vancouver residents in high-amenity areas of the southern coast and the Okanagan in particular denotes an increasingly important aspect of relationship between the primate metropolis and the rest of British Columbia.

The overall picture, however, is unmistakably one of reconfigured core-periphery relationships in British Columbia. As seen in earlier work, economic restructuring has revealed distinct "regional realities" emerging in various parts of the province, influenced by location, factor endowment and comparative advantage (natural or induced), market orientation, and industrial structure, among other factors (Davis and Hutton, 1992). Previous studies have also confirmed increasing *divergence* between Vancouver and the rest of the province in terms of growth rates and cyclical stability (Davis and Hutton, 1989).

A fully elaborated account of the dynamics underlying core-periphery restructuring is manifestly beyond the scope of this chapter, but at the broadest level we can at least identify some of the principal influences. Market factors are certainly important, and indeed it is tempting to ascribe the greatest significance to these: the capacity of corporations to extract and sell resource commodities in world markets in competition with other potential suppliers is, after all, a defining feature of staple economies. In addition, the openness and exposure of staple regions to the vagaries of world markets, and thus to potential change transmitted across vast distances, tends to accelerate transformation within these regions.

But government and policy influences are also significant and may even be decisive in some respects. The early exploration and exploitation of staples in the Canadian periphery by the Hudson's Bay Company was, of course, mandated by royal charter. In the British Columbian case, the configuration of staple economic development was first enabled and then accelerated by public investment in roads, railways, and other infrastructure. The more recent restructuring and reorientation of Vancouver's economy, which represents such a distinct departure from the earlier mode of core-periphery bonding, has certainly been facilitated by a range of public policy decisions at all levels of government. These include, for exemplary purposes, the rezoning of industrial land in Vancouver's core by the City Government, the Provincial Government's sponsorship of Expo '86 (which was seen as "opening" Vancouver to the world), the Federal Government's immigration policy which has *inter alia* provided an infusion of new capital and entrepreneurial energy to Vancouver's economy, and the vigorous international marketing efforts of all three levels of government within the Pacific sphere especially. Whether government can now effectively address the new regional realities in British Columbia, defined largely by the emerging core-periphery dynamics discussed above, and including both an apparently historical shift of growth momentum to the core metropolis, as well as more uncertain prospects in many smaller towns with high resource dependency ratios, will be a fundamental issue for the future.

REFERENCES

Barnes, T.J., Edgington, D.W., Denike, K.G., and McGee, T.G. (1992). Vancouver, the Province and the Pacific Rim. In T. Oke and G. Wynn (eds.), *Vancouver and its region* (pp. 181-200) Vancouver: UBC Press.

Binkley, C.S. (1992). The cross roads in British Columbia's forests. In T.J. Barnes and R. Hayter (eds.), *Troubles in the rainforest: British Collumbia's forest economy in transition* (pp. 13-33). Victoria, B.C.: Western Geographical Press.

City of Vancouver (1986). *Report to Council on the Downtown Eastside Economic Development Society.* Economic Development Division, Department of Finance.

_____ (1991a). *Report to Council on the Proposed Third Runway for Vancouver International Airport.* Joint Finance and Planning Departments Report.

_____ (1991b). *Central Area Plan.* Planning Department.

_____ (1993). *Industrial Lands Review. Part 2: Industrial Activity in the City of Vancouver.* Planning Department.

_____ (1994). *Economic Data Base.* Economic Development Division, Department of Finance.

Coffey, W.J. (1994). *The evolution of Canada's metropolitan economies.* Montreal: Institute for Research on Public Policy.

Conservation of Human Resources (1977). *The corporate headquarters complex in New York City.* New York: Columbia University.

Daniels, P.W. (1993). *Service industries in the world economy.* Oxford: Blackwell.

Davis, H.C. (1993). Is the Metropolitan Vancouver economy uncoupling from the rest of the province? *B.C. Studies,* 98 (summer): 3-19.

Davis, H.C., and Hutton, T.A. (1989). The two economies of British Columbia. *B.C. Studies,* 82: 3-15.

_____ (1991). An empirical analysis of producer service exports from the Vancouver Metropolitan Region. *Canadian Journal of Regional Science,* XIV(3): 371-389.

_____ (1992). *Structural change in the British Columbia economy: Regional diversification and metropolitan transition.* Prepared for the Economy Core Group, British Columbia Round Table on the Environment and the Economy. Victoria, B.C. Round Table.

Hayter, R., and Barnes, T.J. (1992). Labour market segmentation, flexibility and recession: A British Columbia case study. *Environment and Planning C,* 10: 333-353.

Hulchanski, J.D. (1989). *Low Rent Housing in Vancouver's Central Area: Policy and Program Options.* Vancouver, UBC Centre for Human Settlements, for the Central Area Division, City Planning Department.

Hutton, T.A. (1985). *Vancouver: a Profile of Economic Growth and Change.* City of Vancouver Economic Development Division, Department of Finance.

_____ (1994a). Reconstructed production landscapes in the postmodern city: Applied design, intermediate services, and flexible specialization. Paper presented to the Annual conference, Association of American Geographers, San Francisco: 29 III - 02 IV.

_____ (1994b). City profile: Vancouver. *Cities,* II(4): 219-239.

Hutton, T.A., and Ley, D.F. (1987). Location, linkages and labour: The downtown complex of corporate activities in a medium size city. *Economic Geography,* 63: 126-141.

Innis, H.A. (1933). *Problems of staple production in Canada.* Toronto: Ryerson.

Kunin, R., and Knauf, J. (1992). *Skill shifts in our economy: A decade in the life of British Columbia.*

Vancouver: Canada Employment & Immigration, Regional Economic Services Branch.

Ley, D.F. (1992). Gentrification in recession: Social change in six Canadian inner cities. *Urban Geography*, 13(3): 230-256.

Ley, D.F., and Hutton, T.A. (1987). Vancouver's corporate complex and producer services sector: Linkages and divergence within a provincial staples economy. *Regional Studies*, 21: 413-424.

_____ (1991). The service sector and metropolitan development in Canada. In P.W. Daniels (ed.), *Services and metropolitan development: International perspectives* (pp. 173-203). London: Routledge.

Marchak, P. (1995). Globalization of the forest industry. In T.J. Barnes and R. Hayter (eds.), *Troubles in the rainforest: British Columbia's forest economy in transition* (pp. 149-164). Victoria, B.C.: Western Geographical Press.

North, R.N., and Hardwick, W.G. (1992). Vancouver since the Second World War: An economic geography. In T. Oke and G. Wynn (eds.), *Vancouver and its region* (pp. 200-233). Vancouver: UBC Press.

Polèse, M. (1982). Regional demand for business services and inter-regional service flows in a small Canadian region. *Papers of the Regional Science Association*, 50: 151-163.

Resnick, P. (1985). B.C. capitalism and the Empire of the Sun. *B.C. Studies*, 67: 29-46.

Sassen, S. (1994). *Cities in a world economy*. Thousand Oaks, California: Sage.

Scott, A.J. (1988). *Metropolis: From the division of labor to urban form*. Berkeley, CA: University of California Press.

Simmons, J. (1991). Bang Bang Bang—Monitoring the spatial distribution of Canada's financial sector. In M.J. Bannon, L.S. Bourne, and R. Sinclair (eds.), *Urbanization and urban development: Recent trends in a global context*. Dublin: University College.

Vancouver Sun (1995). Lowest-income area caught in condo vice. 14 February: B 12.

Wackernagel, M. (1994). *How big is our ecological footprint? Using the concept of appropriated carrying capacity for measuring sustainability*. UBC: Task Force on Healthy and Sustainable Communities.

Plate 21 (overleaf)   A coastal sawmill ➤

Community Economic Development in a Rapid Growth Setting: A Case Study of Squamish, B.C.

14

Maureen Reed

Department of Geography, University of British Columbia

Alison Gill

Department of Geography, Simon Fraser University

In many parts of Canada, hinterland communities face uncertain futures because of a declining or degraded resource base upon which traditional employment and wealth generation have relied. In addition, government policy initiatives, capital movement and world trade relations, activism by networks of environmental non-governmental organizations (ENGOs), and the affirmation of the rights of First Nations[1] peoples combine in multiple ways to affect the viability of local places in the hinterland. The resulting restructuring of local economies has brought population and employment declines for some resource towns, while others have experienced rapid population and economic growth. For example, some small towns have created new economic opportunities by promoting and developing tourism and/or amenity values. Resource communities near large urban centres have attracted new residents and businesses as the result of displacement from high cost city environments (Bryant, 1994). It is only recently that research about Canada's regional development has focused on locally-induced economic development processes.

As a result of the failure of top-down regional development policies of the 1960s and 1970s, the concept of local development was increasingly discussed and embraced in the 1980s (Decter, 1989; Jean, 1989). Defined as "development which is endogenous and self-centred" (Jean, 1989: 6), local development identifies local opportunities to meet economic objectives such as job creation, employment stabilization, sustenance of tax revenues and appreciation of property values (Filion, 1991). Local initiatives to influence the flow of capital and people to and from communities have emphasized the use of human, institutional, and environmental resources[2] to stimulate increased economic activities.

These initiatives have coincided with senior government interest in bottom-up planning, and the continued demand by local people to become integrally involved in decisions that directly affect them. Consequently, in Canada, new mechanisms for locally-based planning have been introduced, such as citizen advisory committees, local action committees, local resource boards, round tables, and co-management initiatives (e.g., Filion, 1988; Pinkerton, 1989; Go et al., 1992; Bryant, 1993; Landre and Knuth, 1993).

Despite research by economic, rural, and resource geographers, and sociologists and planners, little is known about how citizens in resource communities might be involved in planning for local development. Here, resource communities are defined as those that are predominantly reliant on environmental resources to support locally-based economic activities. This ignorance of local involvement in development planning exists for at least two reasons. First, there is an absence of research on the issue. On one hand, economic geographers have historically focused on metropolitan urban areas (e.g., Coffey and Polèse, 1984; 1985; Cox and Mair, 1988; Duncan and Savage, 1991), or applied regional development models to the local level (e.g., Markusen, 1980). The interest in local development processes of resource towns is only an emergent one (e.g., Filion, 1988; Barnes and Hayter, 1992; 1993). On the other hand, rural geographers and sociologists have focused on the development of agricultural communities in urban fringe regions (e.g., Bryant, 1989) or the characteristics of individual change agents within community settings (e.g., Blahna, 1991). To date, however, they have not linked individual characteristics with local planning institutions.

Second, the dominant paradigm of development in Canadian resource-based communities has been shaped by neo-Marxist interpretations of Harold Innis's staples theory. According to Innis, the Canadian economic landscape was composed of a periphery of scattered resource communities that provided local metropoles with staple commodities for processing and distribution (Barnes, 1993: 358). Following Innis's tradition, other researchers have argued that these communities became underdeveloped in their role as suppliers of labour, raw materials, and capital in the production process and as markets for the products of the more developed regions (e.g., Matthews, 1983; Clement, 1980; Marchak, 1983; Watkins, 1984). While emphasis has been placed on economic processes in shaping regional and community structure, some studies have also explored the implementation of public policy (e.g., Weller, 1980), and transmission of social and cultural values and environmental factors (e.g., Wallace, 1987; Wynn, 1987) in the context of unequal relations between heartland and hinterland regions.

This body of work has argued, in the main, that Canadian resource communities are characterized by their geographic isolation, male-dominated work force, and homogeneous economic structure (e.g., Bradbury, 1988; Bowles, 1992).

Further, these towns are viewed as dependent on outside markets and company goodwill for their viability, and are consequently vulnerable to rapid social change, in the boom-bust cycles associated with resource exploitation (Marchak, 1983). These interpretations give little recognition to the potential for local initiative to influence the nature of economic development within resource communities. What recognition there is for local agency only considers the *reaction* of individuals to external decisions of capital and to forces beyond their control.

At least two cracks are now visible in this paradigm of resource community development. First, a recent review of resource towns in Canada reveals that they are only superficially homogeneous. In fact, these communities are distinctive in a number of ways, including the particular resource sector that dominates economic life, the degree of spatial isolation of the community from other urban centres, gender differences, and the degree of labour force stability (Randall and Ironside, 1996). Economic restructuring has resulted in further differentiation. For example, while economic restructuring has resulted in a shrinking economic base and population for some resource-based communities (e.g., Port Alberni, British Columbia), others face population increases and economic growth in new sectors as they make the transition into a post-staples economy (e.g., Chemainus, British Columbia) (Hutton, 1994).

Second, there is now empirical evidence that, at the local scale, individual actions can have discernible impacts. For example, the success of Chemainus is largely attributed to the initiative and promotion of one person (Barnes and Hayter, 1992). Local municipalities or other local actors may take decisions to influence the patterns of development ranging in scale from individual decisions, such as land purchase, to the establishment of entire planning systems (Bryant, 1994). While some rural geographers such as Cloke and Little (1988) have examined planning within rural areas, they have not discussed in any detail the role of citizen groups and actions that also contribute to decision-making within communities. Given these findings, there is a case to be made for a greater level of interest in the role of local agency and actors to understand local development.

In this context, the purpose of this chapter is to examine the evolution and effects of a community economic development (CED) process in a resource community. Specifically, it will examine a community-based tourism planning process in a resource town that is experiencing rapid growth. The process examined is conceptualized as a community economic development (CED) process, and not local economic development (LED). This is an important distinction and it is discussed more fully in the first part of the chapter. In the second part, the chapter marries sociological characteristics of rapid growth communities with features of local politics. This part emphasizes that CED processes take place within specific socio-political circumstances that must be considered in evaluating the contributions and/or limitations of CED. The chapter uses as a case study the citizen-based tourism planning process in Squamish, British Columbia, and in doing so provides the basis for a more complete understanding of the dynamics of CED in the restructuring of resource communities.

DISTINGUISHING CED FROM LED

Discussing the role of local agency, we distinguish between local economic development (LED) and community economic development (CED) (Table 14.1). The terms CED and LED are frequently used interchangeably or are undifferentiated in the literature on local development (e.g., Barnes and Hayter, 1993; Bryant and Preston, 1987a; 1987b) because both approaches attempt to analyse the efficacy of endogenous development at the local level. Both neoclassical and neo-Marxist interpretations have characterized LED as locally induced economic growth characterized by an increase in real income *per capita*, accompanied by a structural shift in the local economy (Coffey and Polèse, 1984; 1985). At the risk of oversimplifying, our use of the term local economic development refers to a conventional development model which focuses on provision of land and marketing efforts to effect economic growth. Development is determined to a large extent by the decisions of individual private entrepreneurs in the community who make decisions that are primarily market driven. Participatory planning mechanisms are not considered essential features of the local economic development model.

Running parallel to these debates are writers within a community economic development framework who *emphasize* local participation in the establishment of community priorities. At least two features distinguish CED literature from LED literature. First, CED tries to integrate economic activity with other social objectives. According to Bryant (1995),[3] the overall objective of community economic development is to improve the quality of life of community residents, including economic, social and environmental components. Development may be considered as either a positive structural shift in the community's economy (e.g., a significant diversification of the economic base) and/or the putting in place of significant new capacity for positive change (after Douglas, 1989). New capacity may take many forms, such as infrastructure, training and skills development. Development is not limited to economic projects, but rather may include many facets of a local place, including education, health care, recreation and public services (Filion, 1988). Thus, individual initiatives are viewed as means to broader social ends rather than ends in themselves.

Second, CED seeks involvement of, participation by, and accountability to community members on decisions about overall strategy formulation and project implementation. This is quite different from LED, which is derived from market driven decisions of individual private decision makers *in* the community (Douglas, 1989: 29). Local articulation of community goals and values, often articulated as a community vision for development, and active participation of the community in the choice and implementation of development strategies are considered key elements (Bryant and Scarpelli, 1989). Strategic planning is frequently espoused to assist communities in developing long range goals and establishing the capacity to adapt to changing circumstances (Bryant and Preston, 1987b; Filion, 1988). To facilitate effective implementation of CED

Table 14.1 Interpretations of local and community economic development

	LED *NeoMarxist*	CED	LED *Neoclassical*
Purpose	Economic development as growth. Increase in real income per capita.	Integration of economic development with social objectives. Focuses on improved economic circumstances and new 'community capacity' for change.	Economic development as growth. Increase in real income per capita.
	Examines influence of external sources of investment on local chances for survival (e.g., effects of industrial development model).	Emphasis on establishing a community vision within which economic, social, cultural opportunities are evaluated.	Seeks to foster endogenous business development through incentives for local investment and local accountability.
		Seeks long term planning and local self-determination and entrepreneurship.	
Process	Local development considered as a competitive process. Focuses on competition and/or collusion among local elites. Participation in planning. Citizen participation is not usually considered.	Seeks decentralization of control. Seeks broad community input. Emphasizes team building and partnerships across decision making levels and within community.	Seeks control through the individual entrepreneurs. Identifies limited venues for citizen participation.
Locus of control	Extra-local capital or state *or* Local elites, possibly in coalitions with the local state	Decentralized across community stakeholders and other levels of government.	Local entrepreneurs.

initiatives, "collaboration, cooperation and partnership-building" with senior levels of government are promoted (Bryant, 1993: 6).

To summarize, local economic development is primarily concerned with the establishment of economic projects that are generated by individual private decision makers. By contrast, community economic development is primarily concerned with establishing a community vision within which social and economic development objectives are identified. The vision and objectives are achieved through community-based participatory planning processes involving local citizens. Analysts of LED have focused on the participation by local political elites, whereas writers about CED have explicitly considered the role of 'ordinary' citizens in planning for economic changes at the community level. Because they focus on different elements of local development processes, these two approaches might be viewed as opposite poles of a spectrum. However, there is much to be gained from developing a conceptualization of local development that merges the insights of each.

Despite attention to broader-based citizen participation, CED writers tend to assume that a community can be identified with a single and/or common vision, community members are relatively easy to identify, community residents have equal access to political resources, local stakeholders are willing and able to work cooperatively for common ends, and that local efforts are sufficient to bring about desirable development objectives. By contrast, LED writers who adopt a political economy perspective do not assume that a common course of action is necessarily achieved by cooperation. Instead, they point to the diversity of interests held by stakeholders in local processes and identify alternative strategies such as competition and/or collusion (e.g., Cox and Mair, 1988). Thus, those who study LED from the perspective of political economy can contribute a sensitivity to the diversity and nature of citizen intervention at the local level.

A political-economy perspective interprets local development processes as part of a competitive struggle, where people must compete for changing access to environmental and human resources due to economic and social changes (Roberts and Emel, 1992). This interpretation implies that distributions of influence and power must be also shared in planning processes, even at the local level (Ingram et al., 1984; Kearney, 1989; Dale and Lane, 1994). These distributions can be examined through studies on the influence of individuals and institutional structures in shaping local planning processes such as CED. In rapid growth settings, the changing demographic composition and political expressions of interest are entry points into an analysis of changing power relations in the community. These dynamic relations will affect the role and efficacy of CED in influencing the course of local development.

In the following section, a discussion of changing demographics and local politics in rapid growth settings produces a context where different groups have conflicting goals and objectives related to economic development. In this context, CED is viewed as a vehicle by which both the overall vision of the community and the process of decision making is subject to competitive struggles among stakeholder groups.

LOCAL POLITICS IN RAPID GROWTH SETTINGS

Rapid population growth in resource communities undergoing restructuring creates conflict in at least two ways. First, new entrants place individual demands on environmental resources, infrastructure and services. At the local level, new land uses, particularly for housing, may compete directly for land used in other sectors. In turn, demands for housing will alter the biophysical environment and provision of public services. These demands affect the allocative functions of the state, mainly at the local level. Second, and of concern in this chapter, conflict emerges from a broader debate over the appropriate kind and level of growth. New residents are not likely to be dependent on the traditional resource sector, and, thus, introduce new expectations of community life. In studying rapid growth settings, sociologists have long debated whether the interests of newcomers clash with those of long-term residents on issues such as growth management or environmental protection (cf. Molotch, 1976; Ploch, 1978; Price and Clay, 1980; Little and Krannich, 1982; Fortmann and Kusel, 1990). This debate, however, is much more nuanced than this stark dichotomy implies, and varies according to the presiding characteristics and the kinds of economic changes that accompany population growth.

More specifically, some sociological studies have maintained that community leaders, as well as some groups of residents, maintain a strong adherence to the ideology of growth (Molotch, 1976; Little and Krannich, 1982). In particular, local business people whose fortunes are tied to growth and the vitality of the community are considered to be most active in community decision making and policy formation. These people will be supportive of new economic development so long as they perceive economic benefits for themselves and the community. Yet, local businesses that have had long-standing residence will not always benefit from growth as new people to the community frequent new types of establishments and bypass traditional ones. Local chambers of commerce may battle internally over development issues for these reasons (Markusen, 1989). Newcomers may also have a different vision for the community that may be at odds with that of the traditional business sector.

Population growth and diversification through tourism bring new types of jobs and often imply a new local image. No longer are people reliant on environmental resource extraction and/or processing, but rather are reliant on the promotion of the aesthetic appeal of the community and its surroundings. Such an image may pose a threat to the traditional employment base, especially if newcomers attempt to find ways to 'clean up' the local environment.

Thus, a division may appear between those seeking to maintain the *status quo*, or at least to encourage business starts that are consistent with it, and those who seek to change the nature of economic activities in the local community.

Newcomers may not only threaten the substance of economic development, but also the processes by which decisions are made locally. Historically, local economic development in resource communities has been characterized by informal mechanisms used to promote specific economic initiatives. The idea of local dependency (Cox and Mair, 1988; 1989) has been adapted to resource communities (Reed, 1993; 1995) to explain why and how local businesses intervene directly in local economic development processes. The primary interest of locally-dependent firms is in defending and/or enhancing the flow of value through a specific locality. Locally dependent firms may establish coalitions to promote "hegemonic projects" or to oppose new businesses that threaten to compete, and thereby attempt to represent a singular local interest. By relying on traditional, often informal, decision-making processes and connections to local elites, coalitions may try to reduce general public access to the channels of political influence locally. Local municipal governments are susceptible to the efforts of the business sector, because of their dependence on local taxes, and the need to maintain employment and business conditions.

In a rapid growth setting, the impact of new residents may produce cracks in the power structures and open up new venues for citizen participation. Research undertaken in the 1970s and 1980s on 'population turnaround' or 'reverse migration' from metropolitan to non-metropolitan counties in the United States found that a highly skilled, professional class of newcomer became involved in a range of local issues (Ploch, 1978; Rank and Voss, 1982). Newcomers brought new voices, energies, skills and capabilities to conflicts already apparent, and participated heavily in community affairs (e.g., Blahna, 1991; Fortmann and Kusel, 1990; O'Brien and Hassinger, 1992). Whereas long-standing residents were more likely to use informal mechanisms, newcomers brought a formalization of procedures, processes and institutions for decision making (Cortese, 1982).

A CED process is a more formal mechanism for harnessing citizen opinion about development issues. Its introduction, therefore, is not undertaken in a political vacuum. On the contrary, the preceding discussion suggests that the introduction of CED would not necessarily be welcomed if it provides a venue for expressing new, potentially competitive, interests in local development. As a result, attempts to undertake CED in a rapid growth setting are subject to at least three potential outcomes. First, locally dependent firms may be successful in skewing CED processes to their own ends. That is, by lobbying or colluding with traditional power elites, a local business coalition may identify a hegemonic project that would subsume other initiatives or be able to block the endorsement of new undertakings. Essentially an effort to *co-opt* the CED processes, these actions would be consistent with conventional forms of local economic development and the politics of local dependency. Second, locally dependent firms may lose their control over CED and become *marginal* to the planning of CED initiatives. This might occur if the initiatives identified through a CED process were implemented over the objections of traditional power elites. Third, some form of *juxtaposition or integration* may occur between traditional and new models of economic development. These options are examined with respect to a CED initiative in Squamish, British Columbia.

THE CASE STUDY

Squamish, a District Municipality of about 12,000, is located on Highway 99 between Vancouver and Whistler (Figure 14.1). Until the 1980s, Squamish was clearly a 'classical' resource-based community, dependent primarily on the forest industry relying on logging, pulp production, and sawmilling, which in 1981 employed approximately 27 percent of its labour force. This figure does not include log handling activities, Squamish Terminals, or B.C. Rail, which also contributed to the resource economy. By 1991, 18 percent of jobs were held in basic forestry-related jobs. Services had increased their share of the labour force by 20 percent, to account for 47 percent of the labour force.

The economy of Squamish developed relatively separately from Vancouver and Whistler. However, pressures from Vancouver for port facilities, recreational opportunities and affordable housing began to link the fortunes of Squamish and neighbouring municipalities. In addition, Squamish is situated in the southwestern British Columbia tourism region, which has experienced the highest rate of growth in tourism of any region in the province. The resort community of Whistler, in particular, is an international tourist destination, attracting over 1.5 million visitors annually. These factors combined to generate a high rate of population growth. Between 1986 and 1991, the population of Squamish grew, on average, 2.6 percent per year, and between 1991 and 1993, it grew at an average rate of 3.3 percent. At this rate of growth, the population of Squamish doubles in approximately 20 years. Annual residential sales increased between 1990 and 1993 from 177 to 450. In 1990, virtually all sales were in single-detached dwellings, whereas in 1993 approximately one-half were in single-detached, and one-half in attached dwellings, apartments and condominiums. New housing starts have also boomed. In the five years 1989-1993 inclusive, 823 new starts were reported compared to 177 new starts in the preceding five years.

The effects of tourism and local amenity value for new residents have combined to stimulate growth of the community and to diversify its economic base. Squamish is highly accessible to Vancouver and the international market, and

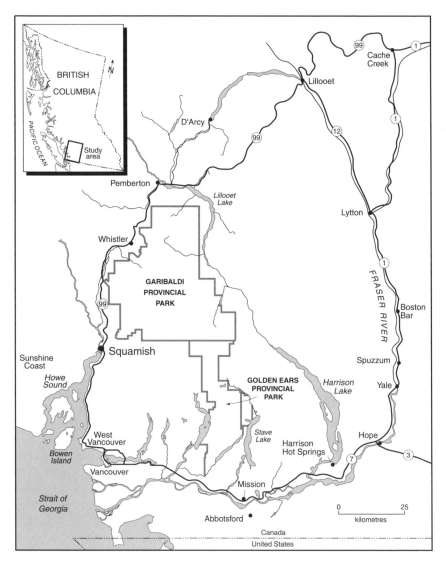

Figure 14.1   Squamish: Location in southwestern British Columbia

offers outstanding recreation opportunities, including rock climbing, windsurf-
ing and hiking. However, the marketing of Squamish as a tourist destination is
hampered by lack of a tourism infrastructure, the nominal aesthetic appeal of
the town centre, restricted access to foreshore areas, and a lack of local
understanding of the current and potential impact of tourism on the local
economy.

In 1992-93, Squamish became involved in community tourism planning through a Provincial government program, the Community Tourism Action Program (CTAP), which provides logistical planning assistance for communities wishing to plan for tourism development. The findings presented here are derived from multiple sources, including involved participant observation during CTAP workshops, interviews with all citizen participants and resource people, and analysis of public documents such as local and regional newspapers, minutes of planning meetings, and government reports. After months of negotiations, the official public launch of CTAP was held in January 1993 with approximately 100 people attending. It became apparent that the initiative to engage in this process did not come from the community at large. Rather, the major incentive by the local council for the initiation of this process was a proposal to develop a four seasons ski development at Brohm Ridge, located on Crown land adjacent to the municipality. The company of Garibaldi Alpen estimated that 80 million dollars of facilities would be created with a total development investment of 360 million by the 11th year. Marketing strategies assumed that, as the nearby Whistler resort nears its development limits, the Brohm Ridge Ski Development would create regional tourism diversity.

Understandably, the proposal was very attractive to some members of the municipal council at the time. Given the potential reduction in forestry jobs in the near future, a ski hill development would offer jobs, taxes, and other local benefits. Also, it would boost the profile of Squamish, providing a launch for a potentially lucrative tourism product. Thus, the municipality and the company lobbied the provincial government for the project. The provincial government declined, stating among other reasons that Squamish Municipality had not 'gone to the people' to determine if this was the type of tourism attraction community residents would like to support. After months of negotiating and posturing, the municipality decided to engage in the CTAP process. For some members of council, the process remained the best means of assuring the provision of the ski hill project. However, as the community-based tourism planning process developed there became an increasing effort by participants to ensure that diverse interests across stakeholder groups were addressed.

THE COMMUNITY TOURISM PLANNING PROCESS

Following the public symposium, volunteers were solicited for a more extensive tourism planning process. Several committees were established. Initially, a tourism coordinating committee was officially formed with representatives from municipal and regional government agencies, the Chamber of Commerce, the Squamish First Nation, and B.C. Rail.[4] It was charged with making recommendations to the municipal council with respect to the formation of a comprehensive tourism plan, the setting of tourism development priorities, the allocation of local resources for assisting tourism opportunities, and the evaluation of ongoing tourism initiatives.

Following the January symposium, approximately 20 community volunteers met to prioritize specific strategies, develop action plan concepts and work on a vision statement to guide the plan. Between February and July of 1993, the volunteers met as two separate advisory committees: one to address process, including promotion, marketing and community education and awareness; the other to address product, including attractions, infrastructure and services. Resource people from the Sea to Sky Economic Development Commission, the Howe Sound Community Futures Society, the Provincial Government agent and the Municipality, were made available to the committees. In July 1993, members from both of the groups combined to form the citizen's advisory committee to develop a vision statement and supporting goals. In addition, a separate committee, the winter tourism development committee, was formed in November 1993 to focus on tourism development specific to the winter months.

Two public meetings were held prior to the finalization of the plan. First, in October 1993 a meeting was organized with a presentation from the tourism coordinating committee as well as Garibaldi Alpen and Associates concerning their proposed Brohm Ridge ski resort proposal. Second, in the absence of sustained participation from the Squamish First Nation, the citizen's advisory committee met with the First Nation people to discuss their tourism vision and to learn about aboriginal issues and projects related to tourism. The ideas raised here were incorporated into the plan. In addition, in an attempt to enlarge citizen input, a draft plan was circulated for comment to a broad group of stakeholders. After several drafts of the plan presented to the tourism coordinating committee, a fourth draft was presented to the District of Squamish Council in June 1994. The plan was approved and adopted by the District of Squamish Council in December 1994 (Citizen's Advisory Committee and Howe Sound Community Futures Society, 1994).

The final plan included a vision statement, 8 tourism goals derived from 103 objectives obtained from the initial public symposium and subsequently ranked by the citizen advisory committee. Of the 103 objectives, 30 were selected to develop into action plan concepts. These concepts were not detailed business plans, but rather were recommendations for future tourism development. No effort was made to identify funding sources, although lead agencies and implementation paths were discussed. Of the 30 concepts, the first 10 related to research, planning, logistical support, training, coordination and infrastructure development. The development of a cruise ship terminal was the top-ranked specific project at number 11. The development of a ski hill was embedded within the third-ranked objective, "to develop a plan to promote outdoor winter tourism opportunities and attractions" (Citizen's Advisory Committee and Howe Sound Community Futures Society 1994, 7).

THE PLANNING EFFORT

The citizen's committee made a distinct attempt to maintain a broad agenda with respect to tourism development, and to include many segments of the community in the planning process. For example, at the first meeting the Brohm Ridge proposal was placed at the top of the agenda, asking the participants to endorse the project. Instead, the committees deferred discussion of this proposal until they had had an opportunity to consider their broader objectives and to hear in more detail the nature, scope and implications of this proposal.

One of the characteristic features of CED processes is the development of a vision statement. It was considered essential to ensure that the goals of tourism development were not simply to attract economic projects, but to address broader issues of quality of life for residents. The phrasing of the statement became one of the most problematic elements of the process. Lack of such a statement ground down the initial attempts to establish a tourism strategy. Crafting a statement took months of hard work by all members. The statement, to "build and strengthen a diverse four season tourism sector while maintaining our small town character and preserving our heritage" (printed on every page of the report) was an obvious attempt to marry new initiatives with the traditional economic base. Heritage was interpreted as cultural and natural heritage, providing a link with the forest sector, the Squamish First Nation, as well as the long-standing Sikh-Canadian population. The concern to maintain the small town character of Squamish had been raised many times in the initial public meeting and was an attempt to ensure that tourism was undertaken in a way that would not compromise the traditional sense of community. The need to include this component was hotly debated, with newcomers being much more insistent on including the term than longer-term residents.

Table 14.2 classifies the main action plan concepts into theme areas and needs. The final report included 30 items. In addition, the winter tourism committee added 6 recommendations to the overall plan, raising the total to 36. Overall, there was support for a broad range of needs from infrastructure, training, community participation, as well as individual attractions and promotions. This suggests that the tourism committees had a clear idea of the range of tasks associated with developing a diversified product. As illustrated in Table 14.2, while the greatest number of recommendations was devoted to developing individual attractions and promotional events, a strong need to undertake research and coordination efforts was identified across almost all categories. Only the winter tourism subcommittee focused most of its effort on identifying particular projects, including an endorsement of the Brohm Ridge proposal. In fact, excluding the work of the winter tourism subcommittee, the research and coordination needs almost balanced those of individual projects.

The committees avoided select recommendations for individual businesses. Participants acted as residents of the community, rather than advocates for

Table 14.2 Classification of action plan concepts of the Squamish tourism development plan

Development Needs	THEME AREAS FOR ACTION PLANS							
	Winter Tourism	Planning & Coordination	Economic & Business Development	Arts, Culture Community Activities	First Nation Heritage & Tourism	Outdoor Adventure	Hospitality, Public Relations Marketing	TOTAL
Attraction/Project	5	1	3	1	0	4	0	14
Promotion/Training	0	0	0	0	0	0	3	3
Infrastructure	1	1	2	0	0	2	0	6
Hospitality/Service	0	0	0	0	0	0	1	1
Research/Coordination	0	1	2	2	2	1	0	8
Community Participation	0	2	0	0	0	0	2	4
TOTAL	6	5	7	3	2	7	6	36

specific interest groups or development projects. Instead, by focusing on research, infrastructure needs and coordination, the emphasis of tourism was shifted from concern for *private project* development to broader *public* concerns about the *process* of development.

MEMBERSHIP ON THE COMMITTEE

The committee was open to community residents willing to participate. Committee members had not previously been part of the local decision making hierarchy, with the exception of one who had, and continues to be, a member of municipal council.[5] Other members included small-scale entrepreneurs, individuals with experience in tourism planning and/or marketing, and concerned citizens. Only one member was a full-time tourism operator. The most distinctive feature of the membership was the length of residence of participants. Three of the 20 volunteers had lived in Squamish for more than 20 years, one had been resident for 10 years. All others had lived in Squamish for 7 years or less, with 11 being resident for 5 years or less.

Of note were those interests not represented. The Chamber of Commerce, a strong voice on the coordinating committee at the outset, did not send a representative to any of the community meetings. Its initial participant held a strong position on the need for a new tourist hotel. In the absence of outright endorsement of this project, the Chamber refrained from further continuous involvement in the committee. In addition, the forest industry was notable by its absence. This gap was noted by members of the advisory committee, yet its deliberations continued without assistance or comment by the forest sector. Involvement of First Nations people was limited. They attended some of the initial meetings, but their attendance and input was not consistent. To ensure that the plan met with the desires and concerns of the First Nations people, the committee held a specific meeting with the local band to discuss the plan. Sikh residents, who comprise approximately 11 percent of the town's population, have historically not participated strongly in community-wide initiatives. While the tourism committees recognized the significance of these people by recommending a cultural festival incorporating Sikh residents, the committee members did not seek direct input from this segment of the community in the plan.

The heightened role of newcomers was prevalent among institutions as well as individuals. Central resource people included employees of the Sea to Sky Economic Development Commission, the Howe Sound Community Futures Society, and the planner in Squamish. These people and their institutions were new to Squamish (5, 2 and 2 years respectively). In particular, the Community Futures Society was responsible for guiding the process and was committed to ensuring that tourism planning would be a community-based process. Without its logistical support, it is not clear if the advisory committee would have completed the plan. At the time that the tourism plan was formulated, advisory committees to municipal council were rare and the municipal government did not have the financial means or staff to offer logistical support for a volunteer effort. Consequently, the planning process marked a shift from traditional power elites and mechanisms (e.g., Chamber of Commerce, Council) to institutional newcomers.

ASSESSMENT OF AND
IMPLICATIONS FOR CED IN SQUAMISH

In this brief case study, the CED process infused new community values into planning for local development. Elements of all three potential outcomes —co-optation, marginality, juxtaposition—were present in this case study. Prior to the CTAP, an individual developer along with some members of the municipal council initially attempted to use the Brohm Ridge proposal as a hegemonic project to define the course of tourism development. Together, they sought local and governmental approvals for the project before the community had a chance to learn and consider its response to the project. The project was subject to protracted lobbying by the municipality to the Provincial government which ended in bitterness on both sides. Broad involvement at the community level had been restricted as specific actors coalesced to reduce the potential for community conflict and to skew participation toward a narrow set of economic values. The council and the developer engaged in the politics of boosterism with the provincial government to generate endorsement and reduce potential conflict. When this failed, the developer turned his lobbying effort towards the citizen advisory committee in the hope of immediate approval of his proposal. These activities were consistent with conventional forms of local economic development and, in particular, the politics of local dependency.

Yet, the proposal was not a priority for the citizen advisory committee. In this context, the creation of the winter tourism committee ensured that the project was not lost from the agenda completely. The committee effectively assured the developer that the project was not marginalized, but would still receive consideration by the community. However, outright co-optation of the process did not occur. Rather, the winter tourism subcommittee acted in parallel fashion to the main advisory committee; its recommendations were only incorporated into the final planning document after some debate. Importantly, although "a" ski hill development was endorsed, no specific project or proponent was identified, and this recommendation was merged with more general winter tourism objectives. Because the recommendation was nested among several others and within a context of a particular vision, it also ensured that the single project did not derail the entire CED process. Instead, the proposal was effectively segregated and then subsumed by the broader-based community effort. If, in the future, the project does not proceed, its failure will not jeopardize the other substantive and procedural elements of the tourism plan. In sum, this series of events represents a juxtaposition of the conventional and the new model of local development.

There were other efforts to realize economic development via the conventional model of local economic development. These efforts had mixed effects. For example, the Chamber of Commerce, a conventional player in local development, withdrew from the tourism planning process when its primary purpose, to obtain approval for a tourist hotel, was not immediately realized. The committee came to its recommendations in the absence of any substantive presence by the Chamber throughout its deliberations. However, the Chamber did retain some authority as it had the opportunity to comment and recommend revisions to the plan prior to its approval by the municipal council. Its major recommendations related to the hiring of individuals to be responsible for tourism coordination and monitoring. In this case, the Chamber did not fear competition among existing businesses, but rather loss of control over tourism planning and promotion in the community. Subsequently, the wording of the recommendations was changed to reflect the need for increased coordination of tourism, rather than the need for a new coordinator. Thus, the Chamber's role was marginal during the planning phase; it did not influence the broad ranging approach taken by participants. As the plan neared completion, however, the Chamber did ensure that its authority was not threatened.

The results of the tourism planning process suggest that the introduction of CED brings new tensions into the traditional decision making processes at the local level. Conventional players in local economic development failed to derail the community-based process entirely, however, they were able to ensure that their basic interests were addressed. Yet, through the commitment of planners and residents in the community, the participatory aims of community development were also beginning to be realized. The plan for the development of tourism is based on a much broader vision of community aspiration than would have been possible by development applications made to council on a project-by-project basis. This result suggests that the narrowly-focused model associated with local economic development and the associated politics of local dependency is not inevitable. The introduction of the community economic development process imposed incremental changes, and coexisted to some degree with a more conventional approach to economic development. The extent to which the CED model will be an influential component of local development processes remains to be seen.

It might be argued that due to the preponderance of new residents, this process was not truly a CED process, but rather, an opportunity to advance a separate hegemonic strategy (e.g., promotion of amenity) by a distinct class of resident. Beyond doubt, those motivated and able to participate included new residents and institutions for whom community-based tourism represented diversification and amplification of amenity values. Yet, most of the members of the committee did not know one another prior to their tenure on the committee. The members were not in agreement about community development; the establishment of the winter tourism subcommittee and the contents of the final document attest to the diversity of positions that individuals brought to the process. In addition, the representatives on the committee, with two exceptions, were not part of the conventional power structure of the community. To consider the final document part of an overall hegemonic strategy is to attribute to the members greater forethought and powers of coercion and influence than they realized in practice.

Notwithstanding these issues, the role of newcomers is clearly influential in realigning the planning process towards new expressions of values locally. The newcomers took advantage of, and worked to establish and shape, new venues for public involvement. The vision for tourism development is a much broader one than the municipal council and the proponent first presented to the residents in the form of a specific ski hill development. It is clear that involvement of newcomers shifted the emphasis of tourism planning from the development of a solely private project towards public goods and services that would be in keeping with community needs and desires. And while it was not entirely successful in obtaining representation from all segments of the local community, there was a much stronger effort to ensure that all segments were considered and/or consulted when the plan was being developed than had been experienced in the municipality previously. Therefore, in Squamish, a resource community being reshaped by rapid growth, CED began to mould both the process and outcome of local decision making.

There remain several outstanding issues with respect to community-based planning in Squamish. There is a need to develop ways to include the concerns and issues of under-represented groups. Those with a direct stake in traditional community characteristics chose not to participate; their influence was thereby minimized. For example, both the conventional sectors, such as forestry, and those that have not had a strong voice in municipal affairs, such as the Sikh residents, were not represented; their voices must be brought into the process if it is to be considered broadly community-based.

In addition, a major hurdle at this stage is to move the plan from a written document to one that guides the implementation of specific initiatives. As the citizen's committee was only of an advisory nature, its efficacy in generating a new focus for local development is yet to be realized. To date, some minor initiatives have been undertaken, such as the development of new signs and the provision of waterfront access to the general public. Thus far, municipal employees have viewed the plan as a document that can be used by the private sector to lend support to private development proposals. The document is plagued by the citation of several government agencies as potential leads in different components. In the absence of an individual leader, commitments by governments and residents for long term funding and support are not obvious. It will take more time to realize whether Squamish will be able to overcome these operational difficulties and to cite whether the CED process has produced tangible results.

There is no doubt, however, that intangibles have been realized. As noted in the definition, CED is also concerned with creating new capacity for change. This capacity may be viewed at the scale of the individual. In this regard, participants believed that their voluntary efforts on the committee had provided several personal benefits. They expressed the benefits of new friendships, a greater sense of community belonging, improved knowledge of community affairs and local power structures, and a willingness to undertake other volunteer duties in the future. One member, who had been involved in a range of volunteer activities, chose to formalize her community service and was elected to municipal council. These intangible results are an important component of CED and are frequently ignored by conventional models of local development.

CHALLENGES FOR UNDERSTANDING CED IN RAPID GROWTH SETTINGS

Despite the potential positive effects of CED, several challenges for rapid growth communities remain. First, the gap between CED planning and implementation continues to plague such initiatives. In Squamish, as the community moves to implement the recommendations of the plan it is difficult to discern overall effects. It is not clear if new community-based institutions will be effective in reshaping the local resource environment. The politics of implementation may reinforce old elites, especially if sectors that did not participate in the planning are able to mobilize and thwart implementation of the plan.

Second, the efficacy of implementation is also questioned. Even if implementation is attempted, it is still unclear if local efforts are sufficient to address the exogenous agents of change and provide for a community-based, diversified tourism package. As Bryant (1995) points out, CED occurs within a multiactor and multi-interest environment. Most decisions that shape the future of communities are made by people, such as householders or local business owners, over whom community "leaders" have little or no control. In a rapid growth setting, these exogenous agents may include the effects of new residents moving in and/or new development opportunities in adjacent municipalities.

Third, the study of rapid change in resource communities is severely hampered by the lack of data at the local level. While census data are available for larger metropolitan areas, for small communities these data are not available or are costly to obtain. It is difficult to determine the extent and effects of rapid change when Statistics Canada data become available at the community level only every 10 years. Other sources of data are scattered, gathered for different purposes, often incomparable, thereby making generalizations speculative. The lack of data places an obvious limitation on the ability of researchers to track changes, but it is also a practical limitation for planners who attempt to deal with rapid change with limited resources.

The final challenge deals with the combination and transformation of conditions and processes at the local level. As discussed earlier, resource communities themselves are heterogeneous according to several characteristics including degree of isolation, type of resource-based activity, and gender roles and relations. Institutionally, communities exhibit diversity and dynamism in terms of community structure, organization, and relations with senior levels of government. As a result of the small scale and unique characteristics of many resource communities, it is not possible to predict local outcomes with any certainty. For example, because of the smallness of scale, the importance of individuals is heightened. Therefore, the experiences in one community may not be automatically transferred to other places. In addition, CED processes themselves are complex. The establishment of CED processes implies the creation, destruction and/or reinforcement of relations within and without individual communities. The results will be processes of varying influence and efficacy over time and across different places. Consequently, the outcomes of CED are likely to be increasingly individualistic and diverse.

If we accept that there is no single model to be sought or studied, how can we draw conclusions about the influence of CED processes in the development of resource towns? In part, our response was to consider CED as a local model of development that can be described in terms of conditions of a specific time and place. This description was used as a heuristic device to assist in examining the structure and organization of different political groups and analysing how they attempted to influence local initiatives and conditions in Squamish, British Columbia. Further research in other localities would be needed to help separate idiosyncratic elements from distinctive ones, and provide specific, if partial, understanding of the efficacy of CED in shaping resource communities.

ENDNOTES

[1]   First Nations relates to the aboriginal peoples who inhabited North America before European occupation. For example, the Mohawk people would refer to themselves as the Mohawk (First) Nation. Reference to "Indians" or "Bands" would reinforce the labels provided by others for these people. We respect the terminology preferred by Aboriginal people in Canada. These people are presently negotiating with Canada to have their rights of occupancy, self-government, and self-determination recognized and affirmed. Affirmation by the governments (federal and provincial) of Canada would alter the structure of authority of lands and associated resources over much of the hinterland areas.

[2]   The terminology of 'environmental resource' is deliberate. It includes components of nature for which economic values are ascribed (e.g., timber) as well as those for which

economic values are difficult to ascertain, despite their importance for aesthetic, scientific, life support benefits or their intrinsic worth (e.g., clean air, nonharvested biotic or wildlife species).

[3]   In his early writing about local development, Bryant (with Preston) called his model "local economic development." Since the late 1980s, however, he has modified his terminology to "community economic development." Given his emphasis on community process and inclusion of multiple stakeholders and interests, his model is consistent with that of "community economic development" described in this chapter.

[4]   This group was unofficially formulated in October 1992 to plan the public symposium. Its membership was made formal in November 1993.

[5]   One member of the tourism advisory committee was subsequently elected to municipal council.

REFERENCES

Barnes, T. (1993). Knowing where you stand: Harold Innis, staple theory, and local models. *Canadian Geographer, 37*, 357-359

Barnes, T., and Hayter, R. (1992). 'The little town that did': Flexible accumulation and community response in Chemainus, B.C. *Regional Studies, 26*, 647-663

Barnes, T., and Hayter, R. (1993). *Economic restructuring and local development at the margin: Forest communities in coastal British Columbia*. Discussion paper, Institute for Northern Studies, Lakehead University, Thunder Bay, Ontario, Canada

Blahna, D.J. (1991). Social bases for resource conflicts in areas of reverse migration. In R.G. Lee, D.R. Field and W.R. Burch (eds.), *Community and forestry: Continuities in the sociology of natural resources* (pp. 159-178). Boulder, Colorado: Westview Press.

Bowles, R. (1992). Single industry resource communities in Canada's north. In D. Hay and G. Basran (eds.), *Rural sociology in Canada* (pp. 63-83). Toronto: Oxford University Press.

Bradbury, J. (1988). Living with boom and bust cycles: New towns on the resource frontier in Canada, 1945-1986. In T.B. Brealey, C.C. Neil and P.W. Newton (eds.), *Resource communities: Settlement and workforce issues* (pp. 3-20). Canberra, Australia: Commonwealth Scientific and Industrial Research Organization.

Bryant, C.R. (1989). *Rural community, land use dynamics and sustainable development*. Keynote address to the International Geographic Union Commission Meeting, Amsterdam, Netherlands.

Bryant, C.R. (1993). *The role of corporate and voluntary sectors as partners in effective initiatives in community economic development*. Presented at the National Symposium on Community Economic Development, Kananaskis, Alberta.

Bryant, C.R. (1994). *The role of local actors in transforming the urban fringe*. Paper presented at the 1994 annual meeting of the Association of American Geographers, San Francisco.

Bryant, C.R. (1995). *Community economic development. Changing the shape of the future through the power of people*. Presented at the Conference Preparing for Now! Community Tools and Strategies for Economic Renewal, Vancouver, Simon Fraser University.

Bryant, C.R., and Preston, R.E. (1987a). *Economic development bulletin number 1: A framework for local initiatives in economic development*, University of Waterloo, Economic Development Program, Waterloo, Ontario.

Bryant, C.R., and Preston, R.E. (1987b). *Economic development bulletin number 2: Strategic planning and local development*, University of Waterloo, Economic Development Program, Waterloo, Ontario.

Bryant, C.R., and Scarpelli, R. (1989). Challenges for the 1990s. In C.R. Bryant and B. Buck (eds.), *Papers in Canadian economic development, Volume 3* (pp. 13-74). Waterloo, Ontario: University of Waterloo and Industrial Developers Association of Canada.

Citizen's Advisory Committee and Howe Sound Community Futures Society (1994). *Tourism development plan for the District of Squamish*. Squamish, B.C.: Citizen's Advisory Committee and Howe Sound Community Futures Society.

Clement, W. (1980). A political economy of regionalism in Canada. In J. Harp and J.R. Hofley (eds.), *Structured inequality in Canada* (pp. 268-284). Scarborough, Ontario: Prentice-Hall of Canada.

Cloke, P., and Little, J. (1988). Public sector agency influence in rural policy making: Examples from Gloucestershire. *Tijdschrift Voor Econ. en Soc. Geografie, 79*, 278-289.

Coffey, W., and Polèse, M. (1984). The concept of local development: A stage model of endogenous regional growth. *Papers, Regional Science Association, 55*, 1-12.

Coffey, W., and Polèse, M. (1985). Local development: Conceptual bases and policy implications. *Regional Studies, 19*, 85-93.

Cortese, C.F. (1982). The impacts of rapid growth on local organizations and community services. In B.A. Weber and R.E. Howell (eds.), *Coping with rapid growth in rural communities* (pp. 115-135). Boulder: Westview Press.

Cox, K.R., and Mair, A. (1988). Locality and community in the politics of local economic development. *Annals of the Association of American Geographers, 78*, 307-325.

Cox, K.R., and Mair, A. (1989). Levels of abstraction in locality studies. *Antipode, 21*, 121-132.

Dale, A.P., and Lane, M.B. (1994). Strategic perspectives analysis: A procedure for participatory and political social impact assessment. *Society and Natural Resources, 7*, 253-267.

Decter, M. (1989). *Diversification and single industry communities: The implications of a CED approach*. LED Paper #10, Economic Council of Canada, Ottawa

Douglas, D. (1989). Community economic development in rural Canada: A critical review. *Plan Canada*, 28-46.

Duncan, S., and Savage, M. (1991). New perspectives on the locality debate. *Environment and Planning A, 23*, 155-164.

Filion, P. (1988). Potentials and weaknesses of strategic community development planning: A Sudbury case study. *Canadian Journal of Regional Science, 11*, 393-411.

Filion, P. (1991). Local economic development as a response to economic transition. *Canadian Journal of Regional Science, 14*, 347-370.

Fortmann, L., and Kusel, J. (1990). New voices, old beliefs: Forest environmentalism among new and long-standing rural residents. *Rural Sociology, 55*, 214-232.

Go, F.M., Milne, D., and Whittles, L.J.R. (1992). Communities as destinations: A marketing taxonomy for the effective implementation of the tourism action plan. *Journal of Travel Research*, Spring, 31-37.

Hutton, T. (1994). *Visions of a 'post-staples' economy: Structural change and adjustment issues in British Columbia*. Planning Paper, Centre for Human Settlements, UBC, Vancouver, B.C.

Ingram, H.M., Mann, D.E., Weatherford, G.D., and Cortner, H.J. (1984). Guidelines for improved institutional analysis in water resources planning. *Water Resources Research, 20*, 323-334.

Jean, B. (1989). Development in its place: Issues in local development. *Canadian Journal of Regional Science, 12*, 5-8.

Kearney, J.F. (1989). Co-management or co-optation?: The ambiguities of lobster fishery management in Southwest Nova Scotia. In E. Pinkerton (ed.), *Co-operative management of local fisheries: New directions for improvement management and community development* (pp. 85-102). Vancouver, B.C.: University of British Columbia Press.

Landre, B.K., and Knuth, B.A. (1993). Success of citizen advisory committees in consensus-based water resources planning in the Great Lakes Basin. *Society and Natural Resources*, 6, 239-257.

Little, R.L., and Krannich, R.S. (1982). Organizing for local control in rapid growth communities. In B.A. Weber and R.E. Howell (eds.), *Coping with rapid growth in rural communities* (pp. 221-241). Boulder: Westview Press.

Marchak, P. (1983). *Green gold: The forest industry in British Columbia*. Vancouver, B.C.: University of British Columbia Press.

Markusen, A.R. (1980). The political economy of rural development: The case of Western U.S. boomtowns. In F.H. Buttel and H. Newby (eds.), *The rural sociology of advanced societies: Critical perspectives* (pp. 405-430). Osmun, Montclair: Allanheld.

Markusen, A.R. (1989). Industrial restructuring and regional politics. In R.A. Beauregard (ed.), *Economic restructuring and political response*, Urban Affairs Annual Reviews, 34 (pp. 115-147). Sage: Newbury Park.

Matthews, R. (1983). *The creation of regional dependency*. Toronto: University of Toronto Press.

Molotch, H. (1976). The city as a growth machine: Toward a political economy of place. *American Journal of Sociology*, 82, 309-332.

Pinkerton, E. (ed.) (1989). *Co-operative management of local fisheries: New directions for improvement management and community development*. Vancouver: University of British Columbia Press.

Ploch, L.A. (1978). The reversal in migration patterns—some rural development consequences. *Rural Sociology*, 43, 293-303.

Price, M.L., and Clay, D.C. (1980). Structural disturbances in rural communities: Some repercussions of the migration turnaround in Michigan. *Rural Sociology*, 45, 591-607.

Randall, J.E., and Ironside, R.G. (1996). Communities on the edge: An economic geography of resource-dependent communities in Canada. *Canadian Geographer*, 40, 17-35.

Rank, M.R., and Voss, P.R. (1982). Patterns of rural community involvement: A comparison of residents and recent immigrants. *Rural Sociology*, 47, 197-219.

Reed, M.G. (1993). Governance of resources in the hinterland: The struggle for local autonomy and control. *Geoforum*, 24, 243-262.

Reed, M.G. (1995). Co-operative management of environmental resources: A case study from Northern Ontario, Canada. *Economic Geography*, 71, 132-149.

Roberts, R., and Emel, J. (1992). Uneven development and the tragedy of the commons: Competing images for nature-society analysis. *Economic Geography*, 68, 249-271.

Wallace, I. (1987). The Canadian Shield: The development of a resource frontier. In L.D. McCann (ed.), *Heartland and hinterland: A geography of Canada*, 2nd edition (pp. 442-481). Scarborough, Ont.: Prentice-Hall Canada.

Watkins, M. (1984). A staple theory of economic growth. Reprinted in W.T. Easterbrook and M.H. Watkins (eds.), *Approaches to Canadian economic history* (pp. 49-73). Ottawa: Carleton University Press.

Weller, G.R. (1980). Resource development in Northern Ontario: A case study in hinterland politics. In O.P. Dwivedi (ed.), *Resources and the environment: Policy perspectives for Canada* (pp. 243-269). Toronto: McClelland and Stewart.

Wynn, G. (1987). The Maritimes: The geography of fragmentation and underdevelopment. In L.D. McCann (ed.), *Heartland and hinterland: A geography of Canada*, 2nd edition (pp. 174-245). Scarborough, Ont.: Prentice-Hall Canada.

Plate 22 (overleaf)   Hauling logs by railway   ➤

Staples, Regional Growth and Community Sustainability

15

G.R. Walter

Centre for Sustainable Regional Development,
University of Victoria

INTRODUCTION: REGIONAL GROWTH AND COMMUNITY SUSTAINABILITY

British Columbia is generally viewed as having a prosperous economy. Nonetheless, there is concern for the economic sustainability of its forestry regions and particularly their communities. For many years these communities have suffered from more or less normal fluctuations in the demand for housing, national or international, but many communities expected that sustained economic activity was assured because of British Columbia's provincial policy of "sustained yield" from forests. For example, the B.C. Forest Resources Commission offered the following:

> The forest resources of British Columbia must provide for the social, spiritual, environmental and economic well being of British Columbians. They must be managed in such a way that these values are maintained for future generations (FRC, 1991: 8).

This is only the latest of many such statements, going back at least to the first 1945 Sloan Commission on forestry in British Columbia (Pearse,1976: D3).

A number of recent forces have shaken confidence in British Columbia forestry based community sustainability: market internationalization, technical change, recognition of the emptiness of sustained yield due to overcutting, shifting views as to the portion of the land base which is appropriate for industrial forestry, and a rising concern for regional ecological integrity. These forces have roots in changing economic structure and natural resource scarcity. It is becoming clear that a sustainable community requires a sustainable regional resource base.

Neither community nor sustainability are easy to define. Community may be a geographical matter, that is, a grouping of individuals living in close proximity. It may be a social matter, consisting of individuals identifying with a common religious, political, or cultural position. It may be an economic matter, based on a communality of economic interests, for example, the financial "community." A given community definition often includes all these and more, and involves overlapping and sometimes conflicting loyalties.

Sustainability is an equally difficult concept. It has received very considerable attention following the United Nations report *Our Common Future* (WCED, 1987). As used here it means the ability of a community to maintain itself over the foreseeable future at an acceptable level of human and ecological welfare. Sustainability includes stewardship and adaptability.

Consequently, community sustainability, as used in this discussion, is a dynamic process which results in the continuous healthy economic and social habitation of a region. Healthy habitation implies that the ecosystem upon which the community depends is not destabilized and the level of human welfare in the community is acceptable. Inasmuch as most of this discussion is directed to economic issues, the focus is on a sustainable economic community.

This chapter considers several frameworks which have been widely discussed as suitable for regional development policy, in the context of appropriate policy criteria. Much of the wider discussion reflects a presumed crisis of large scale, principally corporate, production. This large scale production has come to be called Fordism, and its critics have advanced ideas in a context known as post-Fordism (Amin, 1994). Before turning to the main discussion, however, it is useful to consider the meaning of a central idea, community sustainability.

WHY WORRY ABOUT REGIONAL POLICY AND RESOURCE BASED COMMUNITIES?

For many years it was taken as established that Canadian regional development was driven by the utilization of natural resources: timber, fur, wheat, minerals, forests (Innis, 1930; Easterbrook and Watkins, 1986; McCalla and Huberman, 1994; Hayter and Barnes, 1990). As McCalla and Huberman put it in their introduction, "Canadian economic history was almost completely dominated by a single school, the staples approach." Management of regions and their resources was considered crucial to the health of the nation and its regions. "Staples" theory held that Canada's development could be told as a number of overlapping stories, each portraying a conquering people's struggle with the wilderness and their successful establishment of community life, based on utilization of some locally abundant natural resource. This perspective thought it essential to worry about regions and communities because they were the stuff of national development, and because failure to exploit local resources effectively could squander opportunity or invite disaster.

In the 1960s, however, the consensus view on the benefits of ample natural resources began to break down. The obvious success of some resource poor international economies, most notably Switzerland, Japan, and Singapore, called into question the necessity of a natural resource staple as a motor of growth. The difficulties of some resource rich economies, particularly the natural gas rich Netherlands, whose exchange rate, wages and welfare payments decreased manufacturing competitiveness, were equally troublesome to informed observers (Ellman, 1981). The dutch experience was sufficiently noteworthy that similar cases were characterized as suffering from the "Dutch disease." Those concerned with regional sustainability began to look more deeply at the advantages and disadvantages of natural resource wealth.

One consequence was an intellectual challenge to the staple hypothesis of Canada's national and regional growth. It took some time to work out the theory and provide supporting evidence, but the argument is relatively simple. Let us call this anti-staples position market-individualism because of its emphasis on prices, factor of production adjustments, and commodity-based relationships. The argument requires the existence of two market sectors, natural resource and industrial, facing fixed international or inter-regional prices. If the industrial sector is able to expand to include all available labour, offering a wage generated by its productivity, then the natural resource sector cannot be the motor of increased per capita incomes, although it may be the motor for extensive national growth (Chambers and Gordon, 1966; Green and Urquhart, 1994). National prosperity will then be set by industrial production and trade, not by natural resource riches. Regions and their communities will wax and wane as labour moves in to exploit natural resources or labour moves out to exploit better opportunities in the industrial sector. Since people, under this theory, always have an economic opportunity in industry, natural resources oriented growth and community sustainability are of no essential consequence, although there will be adjustment costs suffered by individuals. Furthermore, government efforts to support regional sustainability may undermine market adjustments, to the disadvantage of all in the long run (Courchene, 1981).

With the arrival of NAFTA and a world market economy, the market-individualism argument has gathered force. Those seeking policies to support sustainability by natural resource conservation or other extra-market means are often viewed as undermining national competitiveness and favouring local selfish interests. Nonetheless, the market-individualist prescription, that those in troubled regions should go peacefully into the night to re-emerge within trade and industry based urban centres, continues to meet considerable and largely successful local resistance.

THREE POLICY FRAMEWORKS

The search for a practical policy framework for regional and community development has been extremely frustrating in British Columbia and elsewhere.

It is useful to review briefly three operational policy frameworks which have been discussed in the literature. They differ in many ways, but major components distinguishing them are their labour processes, their macroeconomic regimes, and their models of regulation (Jessop, 1992). The frameworks are export-base, community development and flexible specialization.

Export-base

As a region of relatively recent settlement, British Columbia shares with other North American and Antipodean nations an economic tension between a frontier, communitarian tradition and a private enterprise, export of natural products, orientation. To achieve contemporary industrialized standards of living, community self-subsistence and reciprocal trading, which characterized aboriginal communities and played a significant role in pioneering communities, have gradually been abandoned in favour of specialized production and the dominance of market trading relations. Economic interaction has become increasingly individualistic and commodified, as have labour processes. From the point of view of contemporary policy, export competitiveness in the production of commodities has become a major theme. Regulation of regional economic life is by the market and the macroeconomic tools of the modern state.

A recent study found that a large number of British Columbia's communities are highly dependent upon exports of primary commodities (staples) to maintain their commodity standard of living (Forest Resources Commission, 1992). In addition, financial inflows based on private investments or transfer payments provide major support to communities in British Columbia. This reality, together with a social mentality that is oriented to market activity, has been a significant influence on economic policy in British Columbia. A result is official policy to create conditions supporting international competitiveness for export industries. This was most apparent in the "restraint" program of Social Credit in the 1980s (Rosenbluth and Schworm, 1984). The tendency is to minimize social, environmental and governmental costs, in contrast to a policy sensitive to the role of both costs and benefits in maximizing Provincial welfare. Even the strategic directions of the New Democratic government of the 90s are heavily weighted to economic strength, economic sustainability and redefining government (B.C. Cabinet Office, 1994).

Community Development

If the export-base concept has had considerable official attention from economic development authorities, community development policy has more often been left to health and social agencies. Recently, however, community development has received more attention from economic development officials, presumably because of a growing recognition that a modern export base is dependent upon human capital and local initiative, in turn, responsive to the socio-economic attractiveness of a community. Attractiveness is rooted in community attitudes, human resources, and institutional and technological infrastructure. There are two major economic variants of community development, community economic development and local economic development. Community economic development has involved promoting good living and working conditions, effective human resource development, and local enterprise. Thus, in British Columbia, community economic development has often focused on the development of entrepreneurship, attraction of investors, and training or retraining of the work force (B.C. Community Economic Development Branch, no date; B.C. Ministry of Industry, and Small Business, 1985; B.C. Ministry of Regional Economic Development, 1990).

Local economic development attempts to find a fit of local interests and skills with local opportunities, market and non-market. It may de-emphasize the external market orientation of the export-base by promoting local institutional arrangements to supplement or supplant the market as a tool of local economic production. Examples are cooperatives and various local economic trading or exchange systems. This local, market augmenting, variant has not had much governmental support in British Columbia (MacDonald, 1991).

Hallmarks of both community development approaches are local problem identification, networking and leadership, supported by enabling legislation and seed funds provided by senior jurisdictions.

Institutionally, the community development approach tends to rely on socially embedded reciprocal labour relations and volunteerism. Non-profit organizations and similar local organizations, operating subject to market constraints and the policies of senior political jurisdictions, are important vehicles for this approach.

Flexible Specialization

The last framework considered here is flexible specialization. Flexible specialization contrasts with the large scale, uniform product, mass production model identified as "Fordism."

Scott and Storper have identified four historical production systems, each retaining relevance to contemporary regional sustainability: the putting-out system, the mill/workshop economy, Fordist mass production, and flexible specialization (Scott and Storper, 1992). The evolution of important natural resource industries in British Columbia, most importantly the forest industry, has moved from a mill/workshop to Fordist export-base economy in the postwar period, and now may be moving to a post-Fordist, corporate, flexible specialization. The Fordist model, which replaced a mill/workshop forest production system in the 1950s and 1960s, features high raw material throughput rates, standard products sold in mass markets, high capital intensity and rigid labour protocol. It is having difficulty adjusting to a much more interactive, small production run, specialized market environment. There is little doubt that many of the current difficulties faced by regional, forest resource based communities are the result of the increasing failure of the Fordist model applied to natural resource extraction (Barnes and Hayter, 1994; Hayter and Barnes, 1990; Travers, 1993).

It is difficult to discover a clear definition of flexible specialization in the literature, but it is clear that the concept encompasses networks of relatively small independent producers as well as networked corporate production, the latter based on profit centres, small production teams, and other decentralized decision making methods. As used here, flexible specialization possesses five major characteristics: 1) relatively small production units (which may be part of a larger corporate unit or not); 2) the ability to adapt production processes and/or product specifications to meet the needs of users of the product; 3) an information and network intensive communication systems which allows individual units to be linked to the larger economy, including formal or informal marketing arrangements; 4) network organized supply and support services which serve the network of flexible producers; and 5) "bottom up" leadership of the network. Thus independent producer's flexible specialization differs from a "putting-out" system in that the latter system is led from the top and does not have a network system among producers, whereas the flexible specialization system has networks led by the producer's themselves; it differs from the mill/workshop model in that it does not rely exclusively on individualistic market relations and informal contacts for its direction, but relies on an information system organized by the network members, as well as market monitoring, and is configured for customized production; and it differs from the Fordist model in that it is not large scale nor mass oriented in its production processes. Corporate flexible specialization has characteristics one through four, but not five, because the corporate network is led from above. In both case, production units are more self-regulating than in the Fordist model, and labour processes are more flexible, individualistic and reciprocal.

THREE POLICY CRITERIA

Each of the frameworks of production noted above, that is, the community development, the market-individualism oriented export-base, and flexible specialization, have advocates who wish to promote an associated regional policy to support their positions. Their advocacy may be based on conservatism, personal experience or reflection on actual regional development experience. Most often, however, policy consideration of these alternative systems is not well founded in criteria for judgement. This section reviews the three production systems using three criteria: Paretian, communitarian, and bioregional.

Economic efficiency is a canon of contemporary neoclassical economic analysis. For neoclassical economists, economic efficiency is judged using a welfare criteria known as Pareto optimality, which is closely associated with exchange efficiency. The Pareto criterion is that economic welfare is maximized when no further exchanges can be made among individuals such that one individual's welfare could be increased and no others decreased. This criteria underlies the market-individualism critique of staples theory noted above. Its diagnosis is that lagging regional economies are the result of poor market adjustment, leading to regionally misplaced factors of production and "inefficiency." The implication is that government policy hindering migration of "surplus" labour to urban industrial centres reduces economic efficiency and regional prosperity. Policy that distorts market signals impedes Pareto-like improvements in individual welfare. In short, the Paretian policy is that the market should rule. Because the market is a social institution for reconciling the preferences and production opportunities of individuals, and because the market can only deal with "private" market goods (those for which it is socially and technically feasible to exclude people from use, and for which use by one person precludes another's use), the economic efficiency criteria is highly individualistic and commodity oriented (a commodity is a good which is divisible and sold in a market). Other values are discounted or ignored. This criterion most directly supports the Fordist export-base model and is relevant to the other cases as well, particularly flexible specialization. Crudely put, this criterion asks: Will a given action lead to a higher (commodity) monetary income for the region?

A communitarian criterion is oriented to the viability of a community as a whole, with due regard for the welfare of the individuals in the community. Whereas a Paretian criterion is rooted in psychological egotism, treating individuals as self-centred and strictly egotistical, the communitarian one assumes a moral community, where *"we ought to give equal consideration to the interest of everyone who will be affected by our conduct"* (Rachels, 1993: 186, emphasis in the original). The concept "think globally, act locally" represents a practical consequence of considering ourselves a moral community. This concept also recognizes the impact of geography and proximity in the formation of community while maintaining connection to the larger community. Regarding economic policy, concepts such as merit goods, basic needs, human development and socio-economic welfare are some of the ideas which reflect a communitarian criterion. Some practical communitarian tools are joint decision making, stakeholder participation, public participation, non-governmental organizations, and networking. As noted above, a major practical problem is defining a community, human, biological, or both, within a given locality. This problem arises, first, because of the many possible dimensions of concern for others, especially if the "others" include non-human sentient beings and, second, because of a lack of theory and information on the scope of the geographic, cultural and economic interaction of individuals. Nonetheless, as a practical matter, especially in the case of natural resource based communities, settlements and their regions are the usual starting point. A communitarian criterion supports the local economic development variant of the community development model, focusing on the socio-economic health of the community, on human resource training, and encouraging local economic empowerment, particularly of entrepreneurs and local exchange. It is consistent with the independent producers variant of flexible specialization if the latter's operation creates local networks of mutual concern, economic or otherwise. It is less supportive of corporate flexible specialization because of leadership from the top. It is largely inconsistent with the Fordist export-base model. This is because of the Fordist model's external commodity trade orientation, featuring standard export products, minimum cost of production goals, and emphasis on market relations. Crudely put, the communitarian criterion is: Does an action improve the long term quality of life for members of a given community?

The last criterion considered here is the bioregional one. A bioregional criterion addresses the nature and functioning of a bioregion. The healthy coexistence and interdependence of human communities with natural communities within an ecosystem is the essence of a bioregional approach. The operative criterion, put simply, is: Does a given action enhance or undermine the functioning of natural ecosystems and their dependent human communities? Practically, bioregions are often identified with watersheds. The bioregional approach differs from the others reviewed here in that it grants, to a greater or lesser degree, objective standing to nature and its functioning. An aspect of the bioregional approach, lacking in the others, is recognition of a healthy natural ecosystem as and explicit goal, with equal or greater standing than economic growth or community stability. This recognition arises instrumentally, viewing healthy human communities as dependent upon and synergistic with healthy natural communities. It also arises ethically, based on the deep ecology concept of the absolute value of nature and of serious limits to trade-offs involving human and natural welfare. A bioregional criterion for regional sustainability judges actions based on biodiversity, ecosystems integrity, endangered species and eco-niches, human relationships to nature, conservation, carrying capacity, and sustained economic yield.

Development processes which are favoured by a bioregional criterion are likely to be more holistic and sensitive to ecological integrity. The bioregional perspective supports smaller scale "appropriate" technology and is suspicious of megaprojects and widespread use of heavy equipment capable of massive land restructuring. It would favour eco-forestry labour processes over industrial ones. In this sense it is non-Fordist in perspective. It is consistent with the local economic development model because of its orientation to local action and local sustainability. It can be consistent with a variant of the flexible specialization model which emphasizes local networking, local adaptive capacity, and a realistic appreciation of local environmental limits.

A bioregional criterion favours integration of regional policy at the watershed level, and is sympathetic to community management of ecologically-based production processes. Macroeconomic tools, such as interest rate policy, impinge as a constraint. Regulatory tools such as environmental impact statements, planning methods, and similar frameworks established by political authorities senior to a region, are required to ensure basic bioregional integrity across bioregions.

This section has introduced three policy frameworks for working toward regional community sustainability. The export-base framework is rooted in an export, market-defined, economic-income-maximization tradition. It demands that a narrow view of "efficiency," based on provision of individuals with market commodities at the cheapest possible cost, be the major criterion of success. It relies on classic macroeconomic and market regulation to achieve short-run monetary wealth. In practice, it largely ignores the contribution of nature to the creation of wealth, as well as spillovers from the economic wealth-creation process that are damaging to natural wealth. In its classic form, it cannot be the basis for achieving regional sustainability, as the work of Innis suggests historically, and as many contemporary economic tragedies attest.

The community development perspective focuses attention on communities as key influences on regional development, and upon the quality of human resources and infrastructure, social and economic, in achieving sustainability. The key goals are adequate economic wealth, community health and sustainability. Unfortunately, this approach is often weakened by implicit acceptance of a market-led export-base framework as the only way to achieve contemporary standards of living. The effect is that the potential of a community perspective to address broader opportunities for sustainable economic welfare is stunted. This can easily lead to a greater emphasis, within a community development program, on entrepreneurial ability than may be warranted, given other economic opportunities. These other opportunities may include cooperative, communitarian, or reciprocal forms of economic organization. The community development model has also been undermined by attempts to fit it into an economic institutional model which remains essentially market and individualistic in character.

Flexible specialization is based on a presumed shift in technology and markets, a shift favouring production processes customized to client's requirements, in contrast with Fordist production. More customized production offers an opportunity for smaller production units, either within a corporate entity or independently, which might be suitable for supporting community and regional sustainability. To achieve community sustainability based on flexible specialization in a modern, increasingly international economy, there are two alternatives. Flexible specialization may be organized either corporately as a "post" Fordist system (Bessant, 1991, 10-11), which may undermine community involvement and control, or it may use networks of local producers to allow bottom-up influence, cooperative input sourcing and product marketing, and quick response to changing marketing conditions (Briss, 1993). If networking is extended to other centres of community concern, in the sense of community development issues, and includes bioregional elements, that is, environmental monitoring and impact assessment, natural resource and ecological monitoring and conservation, and respect for regional carrying capacity, then it may be consistent with the building of sustainable regions. More often, however, community development and bioregional elements are left to chance, such that flexible specialization either becomes a decentralized export-base approach or takes a corporate form. For example, a recent review of the adaptation of three Vancouver Island communities' export-base in the face of Fordist failure emphasized entrepreneurial ability, labour retraining and tourism promotion, as well as the power of the town's major employers, as significant elements of adaptive success or failure (Barnes and Hayter, 1994).

None of the three policy approaches are sufficiently holistic to cope with regional and community sustainability in a world which is becoming critically crowded, such that many economic and ecological systems hitherto taken for granted are breaking down. The fundamental reason for this is that each deals with a sub-domain of the sustainability problem. Each represents a subsystem of the overall ecological-economic-cultural system threatening the sustainability of communities. In the face of a dominant market-individualist paradigm guided by monetary oriented policy indices, for example, based on the sum of market traded goods and services, the necessary systems balance cannot be obtained.

Consequently, the concepts of communitarism and bioregionalism need to be integrated with a realistic appreciation of economic reality, including the wonderful, if ecologically threatening, capacity of market institutions to produce market commodities, in order to gather broad support and achieve a sustainable synthesis.

CONTEMPORARY BRITISH COLUMBIA AND COMMUNITY SUSTAINABILITY

Many British Columbia communities remain directly dependent upon forests or mines, and a few on fisheries. A recent study of the economic base of British Columbia's communities showed about 60 percent to be dependent upon forestry (FRC, 1992). Overall, about 15 percent of British Columbia's employment, directly or indirectly, is estimated to depend on forest activity (B.C. Round Table, 1991, Table 9-1).

Fordist Postwar Development

Most of the discussion regarding regional and community sustainability in British Columbia has been motivated by the adjustment problems of its communities, which find their economic base threatened by changing market, industrial or technical conditions. In particular, under the influence of a market-individualism regime, forestry-based communities moved strongly from a mill/shop model towards a Fordist model. This is indicated by a number of statistics. From 1954 to 1990 the share of the largest 10 controlling companies in timber production rose from 37.2 percent to 69 percent (FRC, 1991b). The number of sawmills dropped from 2,489 in 1955 to 181 in 1989, with an associated increase in capacity per mill from 10.5 tbf to 140 tbf (FRC 1991a). Current survey results show that 26 firms possessed 100 percent of British Columbia's coast and interior pulp capacity, 9 had 100 percent of the paper capacity, and 23 had 100 percent of veneer and plywood capacity (B.C. Ministry of Forests, 1993).

Fordist development of the forest industry in the postwar period was very successful in realizing wealth and high incomes in British Columbia's forest-based communities. Indeed, for many years, British Columbia's Port Alberni region had a very high per capita income, 110 percent of the British Columbia average (Aldridge, 1992). In the heyday of Fordist industry, forest communities and their residents came to expect exceptionally high incomes, and this without exceptional investment in human capital, vigorous local entrepreneurship, or diversification.

That the Fordist policy realized short run economic success is indisputable. However, given current difficulties in maintaining the level of employment and income, one must dispute the sustainability of the policy, and ask to what extent was the prosperity of the past purchased at the expense of the present and future.

It is now clear, based on a number of measures, that the forests of British Columbia have been over-cut. It follows that a portion of the high incomes of forest communities in the past represent the mining of natural capital, not genuine wealth generation. Whether or not a shift to intensive forestry, reinforcing the Fordist model by moving the industry further toward plantation status and using fertilization and intensive silviculture, would provide a high enough sustainable yield to allow sustainable communities at current levels of economic welfare is debatable. Even if possible, it is not likely to support bioregional or communitarian values.

Aside from the sustainable throughput question, another important economic implication of postwar forest sector Fordism needs to be mentioned. One important economic study in the postwar period has suggested British Columbia suffered from a form of the economist's "dutch disease" (Copithorne, 1978), although this viewpoint has been unconvincingly rejected by another study (Percy, 1986). Fordist production processes have spawned a concentrated industry supported by a strong sectoral union movement, which together have possessed sufficient political strength to win very favourable conditions for the use of public resources from the British Columbia provincial government. In particular, the opportunity to utilize public, largely old growth, timber at less than fair market value (to the Provincial landlord) (Travers, 1993: 193) may have resulted in an excessively large high-wage industry, as compared with that which would have emerged if free-market conditions (on the producer, labour and landlord sides) had prevailed. Had this been only a matter of income distribution within the current population, it would be of relatively little concern. But as has already been argued, it has been a matter of income distribution between generations and it has led to wages in British Columbia higher than those that may be consistent with diversification and long run economic health. While high wages seem good, their effect can be to hinder diversification and the development of a broader based economy. This is because other economic sectors have problems competing internationally while paying wages set in part by resource industries that are living on natural capital rather than on genuine economic productivity, and because potentially talented human resources have been induced to spend their efforts in a single, Fordist sector. This is the classic dutch disease where timber rents play the role of natural gas revenues (as in the original Netherlands case).

Fordist development has enriched the postwar generation of British Columbians, but in a manner that is unlikely to be sustainable at the scale to which British Columbia's natural resource-based regions have developed. Judging by the current distress signals from forest-based communities, the facts are beginning to sink in. The opportunities lost by high living based on natural resource wealth cannot be recaptured; foregone human capital and entrepreneurial talent which might have flourished under a different development pattern cannot be regained. On the other hand, British Columbia and its communities are wealthy by almost any standard, even if family incomes are stagnating. Considerable potential exists in the remaining forest and other natural resources base. Does a policy framework based on flexible specialization offer a positive path forward?

Flexible Specialization Again

From the point of view of the three criteria noted in this chapter (Paretian, communitarian, and bioregional), it is likely that the Fordist dominance of the postwar period is becoming increasingly obsolete. Its strong point was its alleged Paretian efficiency, but it is now clear that efficiency was, in part, illusionary because of non-sustainability and less than full cost accounting, and because it very likely has led to an inefficient Provincial macro-economy, one suffering from the "dutch disease." The dominant Fordist forest economy is literally consuming itself.

To recognize that a given development path has stagnated does not automatically imply that an alternative path is desirable. Nonetheless, that alternative is already moving in the direction of flexible specialization. Change is driven by experience with community economic and social development, a clear post-Fordist industrial model of more customized production, and by futurist arguments concerning the potential of computer and information technology for achieving economies of scale, through networking, while maintaining small and dispersed units of production.

From the viewpoint of purely Paretian economic efficiency, it is possible that corporate post-Fordist variants of flexible production may be a competitive or superior model to an independent producer, or community-network variant. Of course, from a communitarian and bioregional perspective, the Fordist model has never been efficient, on the contrary, it is usually viewed as destructive of bioregional and communitarian values.

Of more practical concern, however, is the degree to which historically developed private market and public institutional relationships, for example, logging markets and extraction rights, are adaptable and responsive such that independent producer flexible specialization is a feasible market option for rural areas. An important challenge when judging the economic efficiency of a given system of production is to ensure that the judgement is directed to the necessary purpose, and is made based on a realistic and desirable institutional context.

Is there a realistic context in which small enterprise flexible specialization has thrived? In many very large cities a system of Fordist mass production has coexisted with small business units that have provided specialized, flexible manufacturing and services, featuring various degrees of specialization to meet customer needs. Indeed, one of the strengths of cities is their agglomeration advantages, which provide networks allowing swift personal interaction and inspection of goods or services. This is an example of an historical market form of flexible specialization that continues to this day, a small shop flexible specialization based on a classic market framework, operating in a spatially concentrated manner. The growth of Fordism undermined market-led flexible production. Local economic development supporters now call for a replacement mechanism, market or informatic, to revitalize independent producer flexible production. Examples are found in Oregon, Washington, Denmark, Sweden, Italy and elsewhere (Briss, 1992).

As noted above, a form of flexible specialization is also developing in a post-Fordist mode. As communication and control technology has improved, we have found that Fordist production is moving into a post-Fordist pattern, one of corporate command and control response to a specific consumer or producer's needs, often within a set menu of options, and using networked out-sourcing to smaller enterprises. Examples are the tailoring of a particular automobile's assembly to a customer's menu choices, or the marketing and subcontracting practices of aircraft companies in developing and building large aircraft. This is corporate flexible specialization, based on authoritarian corporate frameworks.

So there are examples of successful flexible specialization in some form, independent producers and corporate. In both cases, there exists an efficient information structure. In the case of independent producers, it is the normal economic market supplemented by personal and informatic networks. In the corporate case, it is an information system internal to the industry or the corporation itself.

CONCLUSION

Flexible specialization may provide new opportunities for community sustainability, provided bioregional and communitarian elements can be built into its institutional structure. When considering the flexible specialization model as a policy tool for resource-based regions in this context, there are a number of issues: Is the resource base in question suitable for a form of flexible specialization compatible with these criteria, or is the scale of production, as often thought for mining, such as to foreclose the possibility? How can the needed information and networking structure be created to take advantage of the opportunities of flexible specialization, in one of its variants, artisan, local entrepreneurial, corporate? Are the political interests in the society open to any needed institutional changes? To discuss these issues in detail would require another forum. It is clear, however, that communitarian and bioregional proponents of re-invigorated resource-based communities in British Columbia see little prospect of the Fordist model being successful. They would view a post-Fordist flexible specialization solution as perhaps effective on a Paretian economic efficiency criteria, but as unable to provide a more dynamic, innovative, bioregional, community-sensitive, and diverse economic outcome.

It is curious that critics of the Fordist evolution of a staple economy, for example, of the postwar British Columbia forest industry, have based their criticism on lack of community control and local entrepreneurial opportunity, and on its elimination of local markets for forest products in favour of a mass production, export-oriented mode of production; whereas the defenders of the Fordist evolution have argued that the system was a "free-enterprise," market one. As Drushka put it,

> *since World War II, some of the loudest advocates of the competitive, or free enterprise, system have been ... corporate bureaucracies ... echoed by governments formed by the Social Credit party ... but very early in the game the people who ran these corporations, along with the government in power, systematically eliminated the basis of the free enterprise system in the forest industry—competitive bidding for access to the timber supply* (Drushka, 1985: 210).

It is as though those who advocate community forest management, community forest enterprises, eco-forestry and diversified forest production, such as furniture wood production, the growing and harvesting of mushrooms, increased diversification based on small-enterprise, and so on, are the true defenders of individual opportunity and enterprise.

Are they taking up Adam Smith's classic argument in modern guise? Smith decried large scale, oligopolistic enterprise. He particularly abhorred the system of government-granted rights growing out of the economics of Elizabethan England, a mercantilist system. His remedy was competitive, market-based free enterprise, a form that can be viewed as flexible specialization using the marketplace as the networking institution. Today, seeing marketplaces dominated by large-scale Fordist production, critics hope that new information technologies will restore small enterprise to competitive status, without the "market failures" which have supported Fordism, and *if* appropriate government policies can be instituted and institutions standing in the way transformed. It's a tall order. It is a long time since anyone has seen true free enterprise or free markets in the British Columbia forest industry. It is not at all clear that a new yeomanry of region and community can be developed in the face of corporate flexible specialization and out and out Fordism, even with encouraging institutional reform.

REFERENCES

Aldridge, D. (1992). *Alberni/Clayoquot/Cowichan Valley community socio-economic adjustment: A statistical review,* Sustainable Communities Initiative Working Paper No. 6, Centre for Sustainable Regional Development, University of Victoria, Victoria, B.C.

Amin, A. (ed.) (1994). *Post-Fordism, a reader.* Oxford, UK and Cambridge, USA: Backwell.

Barnes, T., and Hayter, R. (1994). Economic restructuring, local development and resource towns: Forest communities and coastal British Columbia. *Canadian Journal of Regional Science,* XVII, 269-310.

Barker, T., and Brailovsky, V. (1981). *Oil or industry? Energy, industrialization and economic policy in Canada, Mexico, the Netherlands, Norway and the United Kingdom.* London: Academic Press.

Bessant, J. (1991). *Managing advanced manufacturing technology: The challenge of the Fifth Wave.* Manchester and Oxford: Blackwell.

Bliss, II. (1993). *Making wood work: Value added policies and programs.* Pacific Northwest Economic Region, University of Washington, Seattle.

B.C. Cabinet Office (1994). *The Plan, Strategic Priories 1994/1995,* 1-18. Victoria, B.C.

B.C. Ministry of Forests (1993). Major primary timber processing facilities in British Columbia. Economics and Trade Branch. Victoria, B.C.

B.C. Round Table on the Environment and the Economy (1991). *The structure of the British Columbia economy: A land use perspective.* A study prepared by the B.C. Ministry of Finance and Corporate Relations, Planning and Statistics Division, March, 1991. Victoria, B.C.

B.C. Community Economic Development Branch, Ministry of Development, Trade and Tourism (no date). *Community economic development manual—Operations manual for economic development committees.* Victoria, B.C.

B.C. Ministry of Industry and Small Business (1985). *Community organizations for economic development programs and economic strategy planning grants program,* 1-23. Victoria, B.C.

B.C. Ministry of Regional and Economic Development (1990). *COED: Community Organizations for Economic Development, Guidelines,* 1-15. Victoria, B.C.

Broadbead, D. (1994). Community economic development practice in Canada. In B. Galaway and J. Hudson (eds.), *Community economic development: Perspectives on research and policy* (pp. 2-12). Toronto: Thompson Educational Publishing.

Copithorne, L. (1978). *Natural resources and regional disparities: A skeptic's view.* Discussion paper No. 106, Economic Council of Canada, Ottawa.

Chambers, E., and Gordon, D. (1966). Primary products and economic growth: An empirical measurement. *Journal of Political Economy,* LXXIV(4), 315-332.

Courchene, T.J. (1981). A market perspective on regional disparities. *Canadian Public Policy,* VII(4), 506-518.

Drushka, K. (1985). *Stumped: The forest industry in transition.* Vancouver/Toronto: Douglas and McIntyre.

Drushka, K., Nixon, B., and Travers, R. (1993). *Touch wood: BC forests at the crossroads.* Madiera Park: Harbour Publishers.

Ellman, M. (1981). Natural gas, restructuring and re-industrialisation: The Dutch experience of industrial policy. In T. Barker and V. Barilovsky (eds.), *Oil or industry? Energy, industrialization and economic policy in Canada, Mexico, the Netherlands, Norway and the United Kingdom* (Chapter 6). London: Academic Press.

Ernste, H., and Meier, V. (eds.) (1992). *Regional development and contemporary industrial response: Extending flexible production.* London and New York: Bellhaven Press.

Easterbrook, W.T., and Watkins, M.H. (1986). *Approaches to Canadian history.* Ottawa: Carleton University Press.

FRC (B.C. Forest Resources Commission) (1991a). Managing British Columbia's forest resources: Structural alternatives, financial considerations, and valuation estimates. *Background Papers - Vol. 4,* Victoria, B.C., FRC, 1-64, tables and attachments.

FRC (B.C. Forest Resources Commission) (1991b). *The future of our forests.* Victoria, B.C., FRC, 1-74.

FRC (B.C. Forest Resources Commission) (1992). *British Columbia community employment dependencies.* Victoria, B.C., FRC, 1-23, A1-A16.

Green, A.G., and Urquhart, M.C. (1994). National development, 1870-1926. In D. McCalla and M. Huberman (eds.), *Perspectives on Canadian economic history* (pp. 158-175). Mississauga, ON: Copp Clarke Pitman.

Hayter, R., and Barnes, T. (1990). Innis, staple theory, exports and recession, British Columbia 1981-1986. *Economic Geography,* 166, 156-173.

Innis, H. (1962). *The cod fisheries: The history of an international economy.* New Haven: Yale University Press.

_____ (1930). *The fur trade in Canada: An introduction to Canadian economic history.* New Haven: Yale University Press.

_____ (1946). *The political economy of the state.* Toronto: Ryerson Press.

Jessop, B. (1992). Post-Fordism and flexible specialization: Incommensurable, contradictory, complementary, or just plain different perspectives? In H. Ernste and V. Meier (eds.), *Regional development and contemporary industrial response: Extending flexible production* (Chapter 2). London and New York: Bellhaven Press.

MacDonald, D.R.B. (1991). *Provincial policy and local initiative: Community economic development in Nanaimo.* Master of Art Thesis, Department of Political Science, Univeristy of Victoria.

Patterson, G. (1990). *History and communications: Harold Innis, Marshall McLuhan, and the interpretation of history.* Toronto: University of Toronto Press.

Pearse, P. (1976). *Timber rights and forest policy in British Columbia, Volume 2*, Appendix D. Victoria, B.C.

Percy, M. (1986). *Forest management and economic growth in British Columbia.* Ottawa: Economic Council of Canada.

Rachels, J. (1993). *The elements of moral philosophy*, 2nd ed. New York: McGraw-Hill.

Redish, A. (1985). An analysis of the 1985/1986 British Columbia budget, *Paper P-85-03.* Vancouver: Pacific Group for Policy Alternatives.

Rosenbluth, G., and Schworm, W. (1984). The new priorities of the Social Credit government, *Paper P-84-3.* Vancouver: Pacific Group for Policy Alternatives.

Scott, A.J., and Storper, M. (1992). Regional development reconsidered. In H. Ernste and V. Meier (eds.), *Regional development and contemporary industrial response: Extending flexible production* (pp. 3-24). London and New York: Bellhaven Press.

Travers, O.R. (1993). Forest policy: Rhetoric and reality. In K. Drushka, B. Nixon, and R. Travers (eds.), *Touch wood: BC forests at the crossroads* (pp. 171-224). Madiera Park: Harbour Publishers.

WCED (World Commission on Employment and Development) (1987). *Our common future.* Oxford: Oxford University Press.

Contributors

Trevor J. Barnes is a Professor in the Geography Department, University of British Columbia.

Clark Binkley is Dean of the Faculty of Forestry, University of British Columbia.

Bill Cafferata is the Chief Forester for MacMillan Bloedel.

Otto Forgacs is now retired, but formerly he was Vice President for Research and Development, MacMillan-Bloedel.

Alison Gill is Chair and an Associate Professor in the Geography Department, Simon Fraser University.

Tom Gunton is an Associate Professor in the School of Resource and Environmental Management, Simon Fraser University.

R. Cole Harris is a Professor in the Geography Department, University of British Columbia.

Roger Hayter is a Professor in the Geography Department, Simon Fraser University.

Tom Hutton is an Associate Professor in the School of Community and Regional Planning, University of British Columbia.

Patricia Marchak is a Professor in the Department of Anthropology and Sociology, University of British Columbia.

Michael M'Gonicle is Professor and Eco-Research Chair, University of Victoria.

Maureen Reed is an Assistant Professor in the Geography Department, University of British Columbia.

Gerald Walter is Professor in the Economics Department, University of Victoria.

Bruce Wilkinson is Professor in the Economics Department, University of Alberta.

Bruce Willems-Braun holds a SSHRC post-doctoral fellowship at the University of California at Berkeley.

Jeremy Wilson is Professor in the Political Science Department, University of Victoria.

Do